Mobile Penetration Testing

As mobile applications become prime targets for cyber threats, *Mobile Penetration Testing: A Hands-On Approach* provides a comprehensive, practical guide to assessing and securing Android applications. This book follows industry-standard methodologies, covering key topics such as Android architecture, OWASP Mobile Top 10, reversing, traffic interception, and real-world hacking scenarios.

Through step-by-step walkthroughs, case studies, and hands-on exercises, readers will learn to identify and exploit vulnerabilities effectively. The book is suitable for cybersecurity professionals, students, and enthusiasts and also serves as a valuable resource for those preparing for certifications like eMAPT and others. If you're a beginner, this guide equips you with the skills needed to perform mobile penetration testing in real-world environments.

Mobile Penetration Testing

A Hands-On Approach

Sabyasachi Paul and Dr. Sajal Saha

CRC Press is an imprint of the
Taylor & Francis Group, an **informa** business

Designed cover image: Shutterstock Image

First edition published 2026
by CRC Press
2385 NW Executive Center Drive, Suite 320, Boca Raton FL 33431

and by CRC Press
4 Park Square, Milton Park, Abingdon, Oxon, OX14 4RN

CRC Press is an imprint of Taylor & Francis Group, LLC

ISBN: 978-1-032-58515-4 (hbk)
ISBN: 978-1-032-58517-8 (pbk)
ISBN: 978-1-003-45047-4 (ebk)

DOI: 10.1201/9781003450474

Typeset in Sabon
by Apex CoVantage, LLC

Sabyasachi Paul
Dedicated to Mom and Dad, for patiently listening to my struggles, and to Lord Bhairava, for relentlessly fueling the fire within me.

Dr. Sajal Saha
Dedicated to Prof. Samit Ray, Chancellor, Adamas University – a visionary academician committed to his mission of serving society through education and innovation.

Contents

Preface

The field of cybersecurity is evolving at an unprecedented pace, with mobile applications becoming a prime target for security threats. *Mobile Penetration Testing: A Hands-On Approach* is designed to bridge the gap between theory and practice, offering a comprehensive, hands-on guide to assessing the security of Android applications.

This book aims to equip cybersecurity professionals, students, and enthusiasts with practical skills to perform Android penetration testing, following industry-standard methodologies. By providing step-by-step walkthroughs, real-world case studies, and hands-on exercises, this book enables readers to identify vulnerabilities, exploit security flaws, and secure mobile applications effectively. It also serves as a valuable reference for certifications such as eMAPT (eLearnSecurity Mobile Application Penetration Tester) and other mobile pentesting certifications.

With a structured approach, the book covers essential topics like Android architecture, OWASP Mobile Top 10, reversing and traffic interception, tool setup, and practical hacking scenarios using real-world vulnerable applications. Our goal is to ensure that both beginners and experienced professionals can apply these techniques confidently in real-world cybersecurity environments.

We extend our gratitude to Adamas University for providing the necessary infrastructure and support and to our readers, whose passion for cybersecurity continues to inspire us. We hope this book serves as a practical and insightful resource in your journey toward mastering mobile application security.

Acknowledgments

The author, Sabyasachi Paul, would like to express his sincere gratitude to Professor Sajal Saha, his mentor during his undergraduate studies, for his invaluable guidance and support. Professor Sajal Saha, as a coauthor, extends his heartfelt appreciation to his family – especially his parents, mother-in-law, wife Samata, and little daughter Adrika – for their unwavering support and understanding. Their patience and encouragement have been invaluable, particularly during the prolonged periods of research when he was unable to fulfill his family responsibilities.

Both authors are deeply grateful to the Department of Computer Science and Engineering, Adamas University, for providing the necessary support, infrastructure, and a conducive research environment, which greatly contributed to the successful completion of this work.

A special acknowledgment goes to the publishers for their unwavering support and commitment to ensuring the highest quality of this publication. Their meticulous editing, formatting, and timely follow-ups played a crucial role in presenting this book in its best form.

Both authors also extend their appreciation to the students of the CSE Department, Adamas University, whose curiosity and enthusiasm for learning have always been an inspiration. We hope this book serves as a valuable resource for students, researchers, and professionals in the field of computer science.

Acknowledgements

About the authors

Sabyasachi Paul is a cybersecurity professional specializing in penetration testing, offensive security, network security and engineering, and digital forensics. With expertise in offensive security, focused in web applications, mobile applications, internal and external network pentesting, multi-cloud pentesting (specializing in AWS – Amazon Web Services and GCP – Google Cloud Platform), and OSINT methodologies, he has earned multiple industry certifications, including eJPT, MCRTA, CAP, CNSP, CBBH, and CCIO. Currently working as a Network and Cloud Deception Engineer at Zscaler, Sabyasachi has hands-on experience in offensive security, threat detection, deception technologies, and securing enterprise environments. Passionate about hacking and red teaming, he is dedicated to sharing practical knowledge with aspiring cybersecurity professionals.

Dr. Sajal Saha is a distinguished academician with more than 21 years of experience in higher education, spanning both affiliated colleges and universities. He currently serves as Professor and Head of the Department of CSE and Director of Product & Innovation at Adamas University. Previously, he was Principal of Meghnad Saha Institute of Technology under Techno India Group. He has also held leadership roles at Kaziranga University, including Dean of Computing Sciences, Dean of Research, and Associate Dean (Compliance). Dr. Saha earned his PhD from NIT Durgapur and has made significant contributions to research, publishing 50 papers in prestigious international and national journals and conferences. He has authored three books and successfully led funded projects worth over INR 60 lakhs from AICTE (All India Council of Technical Education), ASTEC (Assam Science Technology and Environmental Council), ICSSR (Indian Council of Social Science and Research), and the Institute Innovation Council. Additionally, he holds three granted patents and two registered copyrights for his work. Dr. Saha has been actively involved in mentoring students, including guiding the winning team of Smart India Hackathon 2024. His commitment to academic excellence, research, and innovation continues to make a significant impact in the field of computer science and engineering.

Abbreviations

adb	Android Debug Bridge
AOT	Ahead of Time
API	Application Programming Interface
APK	Android Package Kit
ART	Android Runtime
AVD	Android Virtual Device
AM	Activity Manager
CERT	Certificate
CI	Continuous Integration
CLI	Command-Line Interface
CPU	Central Processing Unit
DEX	Dalvik Executable
DVM	Dalvik Virtual Machine
GIAC	Global Information Assurance Certification
GPS	Global Positioning System
GUI	Graphical User Interface
HTTP	Hypertext Transfer Protocol
IDE	Integrated Development Environment
IP	Internet Protocol
ISAC	Information Sharing and Analysis Center
JDK	Java Development Kit
JIT	Just-In-Time
JVM	Java Virtual Machine
MF	Manifest File
MSTG	Mobile Security Testing Guide
OWASP	Open Web Application Security Project
PCAP	Packet Capture File
PM	Package Manager
RAM	Random Access Memory
RSA	Rivest-Shamir-Adleman (Encryption Algorithm)
SDK	Software Development Kit
SEAndroid	Security-Enhanced Android
SQL	Structured Query Language

SQLi	SQL Injection
TCP	Transmission Control Protocol
TLS	Transport Layer Security
UDP	User Datagram Protocol
UAC	User Account Control
UID	User ID
VM	Virtual Machine
VMX	Virtual Machine Executable
XML	Extensible Markup Language

Chapter 1

The essentials

Android phones are so popular that they are being used across the globe. They are in the hands of everyone, often in pairs. Android phones are a great way to pass the time. Have you ever been curious about the components of this tiny computer? What is the compatibility level of your chosen social media apps with the platform? How can you be notified whenever your phone's battery life drops below a certain percentage (preferably 20%)? So, as an aspiring Android penetration tester, you should already know the responses to these questions. Without a doubt, Android penetration testing is one of the most exciting areas to focus on in the field of cybersecurity.

This chapter will assist you in building a strong foundational understanding of how Android devices operate on a software level. The architecture of Android is explained in detail in this chapter, along with other important details that a pentester should be aware of. The prerequisites are a basic understanding of virtualization, the Linux command line, and programming (ideally in Java). So, let's get going and investigate the field of Android pentesting.

1.1 ANDROID ARCHITECTURE

The Linux kernel is located at the fundamental base of the Android architecture. Above the kernel layer are the Libraries layer and the Android Runtime (ART) layer. Above that present is the Application Framework, and the Applications Layer is at the very top. Let's examine these levels in more detail.

Android is based on the Linux platform, and the first layer to interact directly with the hardware is the Linux kernel. It provides fundamental software needed for booting, power and memory management, device drivers, process management, etc. It is the heart of the Android architecture. It manages all the drivers that are required during the runtime. There are several features of the Linux kernel such as it can handle the security between the application and the system. It can efficiently handle memory management and process management, like allocating resources to processes and so on. It is also sometimes referred to as the kernel layer.

DOI: 10.1201/9781003450474-1

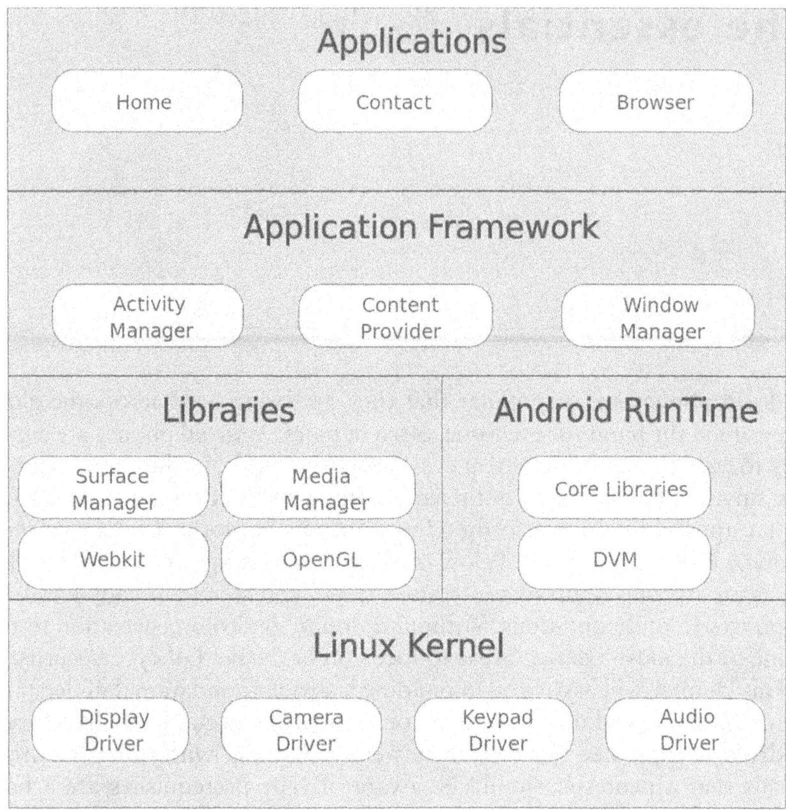

Figure 1.1 Android architecture.

On top of the Linux kernel lie libraries and ART. The libraries include many C/C++ core libraries and some Java libraries to help with Android development. Some libraries to mention are SGL & OpenGL, SQLite, WebKit, etc. The libraries in Android act as a transaction layer between the Linux kernel and the application framework. It provides common services available for the applications to run and execute.

Next to libraries in the same layer comes the ART, which is the shortened form of Android Runtime. The ART comprises the DVM (Dalvik Virtual Machine) (we will understand in the later phase of this chapter) and some core libraries. As of now, consider DVM as a space where our app runs. The core libraries in the ART interact with the application and help it run. The ART provides support for the application framework and powers our application with the help of the core libraries. This layer is often known as the Middleware layer.

Above the ART and Libraries lies the Application Framework. It supports the applications to carry out tasks of different types. It contains several classes that provide supports to several components of the applications. To make it simple, consider your phone's battery drops to 15%, so your phone must give a notification to you. Therefore, the notification manager in Android is a class which will provide support for the app to broadcast a notification to the user. Similarly, for an application to perform various tasks, it needs various support from Android internals. To make it easy, you can consider classes present in the application layer as the drivers required for your app to run. Likewise, for a new hardware device you require a driver; similarly, classes are drivers for Android apps. This will become clearer to you when we touch upon the Android components. This layer is also known as the Framework Layer.

Above all lies the application layer where all the built-in apps are installed, such as Phonebook, Camera, and Gallery. This is the layer with which users interact, like scrolling social media most of the time and watching stories and reels posted by other end users like us.

The Android architecture is developed in such a way that every time the next layer will support its previous or top layer. Now let's take a look at the DVM part.

1.2 DALVIK VIRTUAL MACHINE

All the applications installed on our phone run in a separate virtual environment, though for a specific purpose, which is performance and not security, which many of you might think, and that virtual environment is what the

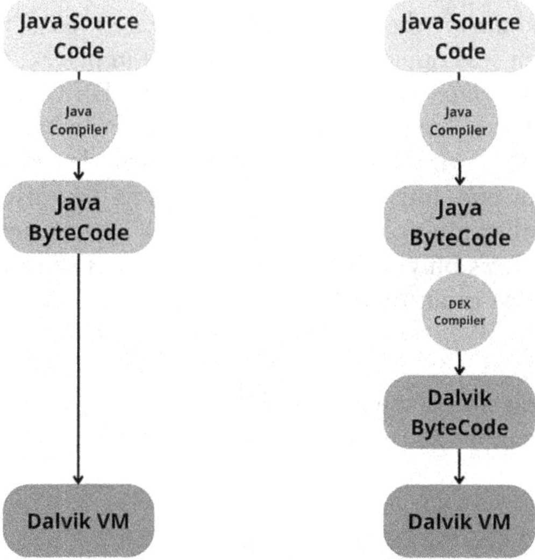

Figure 1.2 Working of DVM and JVM.

DVM provides. The DVM is a virtual machine, which is a virtual space or environment on its own, is a virtual ecosystem where each app runs and performs its task and no other apps has the permission to communicate with each other. We'll discuss briefly the Android sandboxing model later in this chapter, but first let's understand how a Java program works.

Let us suppose that we are trying to print "Hello World" in Java, the .java file gets compiled and generates a .class file, which is also known as the Java bytecode. Now the bytecode ends up in the JVM (Java Virtual Machine). The JVM understands it as a Java bytecode, and by using a JIT (Just-In-Time) compiler, it converts the bytecode into the machine code and then sends it to the CPU to execute it. Therefore, as a result, we get to see Hello World on the screen. Similarly, in the case of Android app development, developers use Java to develop apps. But there are other languages out there in the market through which you can develop your app, like Kotlin and Flutter, to name a few. In that case, the Kotlin compiler will compile the app and generate the same .class file which is a Java-compatible bytecode. After the bytecode is generated, it goes to the DEX (Dalvik Executables) compiler and generates the .dex file, also known as the Dalvik bytecode, which runs on the DVM.

In the later versions of Android, the DVM got replaced with ART. While DVM uses the JIT compiler, ART uses the AOT (Ahead of Time) compiler. Now, to understand what JIT and AOT are, JIT is a type of compiler that compiles during the execution of the program, which means it compiles when the user opens the app. However, AOT is a type of compiler that compiles the app during its installation.

ART was introduced in Android as a replacement for DVM because it uses the AOT compiler. Now, since ART uses AOT, all the compilation is done beforehand during the installation, and hence it leads to faster loading of the app, which was opposite when DVM was in play.

Now, to differentiate between DVM and JVM, DVM is actually register based, specifically designed to run on low memory, whereas JVM is stack based. DVM uses its very own bytecode and runs .dex files, whereas JVM uses a Java bytecode and runs .class files. DVM has also been designed in such a way that multiple instances of the VM can run on the device, whereas JVM shares only a single instance. DVM is supported in Android only, whereas JVM is out there for Windows, Linux, and even Mac. Finally, the executable in DVM are .apk files, whereas in JVM the executable are .jar files.

Now, let's understand some Android components, which is beneficial to know before we move ahead with Android pentesting.

1.3 ANDROID COMPONENTS

Android components are the essential building blocks of an application. These components are loosely connected or attached to the AndroidManifest.

xml file, about which we will study later. In android, there are several components. Let's have a look at those:

Activity: Just like a webpage to a website, a similar activity is to an app. These are the screens with which the user interacts while using an app.
Broadcast receiver: Notifications received on the phone are called broadcasts, and the component which is responsible for passing the broadcast to the user is the broadcast receiver.
Service: Long-running tasks that require no UI and have to be explicitly stopped by the user itself are known as services. They are especially the tasks that run in the background, and you have nothing to do with them directly, like Bluetooth and GPS, to name a few.
Content Provider: Component present in an app that provides data to another app. For example, WhatsApp is asking for an image from the Gallery app, and the content provider present in the Gallery app will provide the required data to WhatsApp.
Content Resolver: Component present in an app that asks for data from another app. For example, WhatsApp requests an image from the Gallery app, so the content resolver of the WhatsApp app will make the request to the content provider of the Gallery app. As discussed above, the content provider of the Gallery app will serve the request being made by the content resolver of the WhatsApp app.
Intent: The component helps in communicating between activities. For example, in the Amazon app, we are searching for a product to buy, you clicked on the Buy Now button, and the screen will be redirected to the payment and checkout activity. This communication between activities, which decides which activity to load based upon certain events, is decided by the Intents.

These intents are of two types:

a. **Explicit:** intent that communicates between activities of the similar app, for example, the shopping app, which switches the screen from product listing to shipping.
b. **Implicit:** intent that communicates between activities of different apps, for example, a mail containing a YouTube video link. So, you are going to open the mail through your mailing app, and the link you will click will open the YouTube app. This switching of activity, which involves another app, is done by the implicit intent.

Intent Resolution: The component which checks which implicit intent needs to call which activity, for example, a mail you received having links to Twitter, Facebook, and YouTube. So, clicking on different links will result in the popping up of different activities. Therefore, this check is done by intent resolution.

Great, after brushing up on the basics so far, let's meet our friend APK, and let's see what it has to offer.

1.4 ANDROID PACKAGE KIT

We all have used Windows at some point in our lives and might also be familiar with software, like Word, PowerPoint, and Excel. Remember the file extension each file was having? .xlsx for Excel, .docx for Word, .pptx for PowerPoint, etc. Similarly, Android apps also have extensions .apk. APK stands for Android Package Kit. It is similar to a zip file bundled with source codes, images, font files, styling files, etc., but in an encrypted form. A simple *unzip* command in the Linux terminal would extract all the files from an apk, but it would be encrypted. There we use certain tools and softwares like Jadx, which we'll look at in Chapter 6, which helps us to read the encrypted source code of the apk.

As per Figure 1.3, a simple unzip command extracted all the files and folders that came bundled together in an apk. Let's visit all the files and folders and understand what role each of them has.

AndroidManifest.xml: An XML (Extensible Markup Language) file that contains all the permissions, activities (screens), content providers, etc. required for the app to run is listed. It is encrypted since we only unzipped it and did not decrypt. It contains all the application configuration information and defined security parameters.

classes.dex: It contains the Java bytecode in a .dex format. This dex file can be decompiled, and the application source code can be read. This file will actually run on DVM.

```
┌──(kali㉿kali)-[~/Desktop/app]
└─$ unzip -q diva-beta.apk

┌──(kali㉿kali)-[~/Desktop/app]
└─$ ls
AndroidManifest.xml  classes.dex  diva-beta.apk  lib  META-INF  res  resources.arsc

┌──(kali㉿kali)-[~/Desktop/app]
└─$ ▮
```

Figure 1.3 Inside of an APK.

```
┌──(sabyasachi㉿DESKTOP-JFB26PG)-[/mnt/c/Users/sabya/3D Objects/diva]
└─$ unzip -q diva-beta.apk

┌──(sabyasachi㉿DESKTOP-JFB26PG)-[/mnt/c/Users/sabya/3D Objects/diva]
└─$ ls
AndroidManifest.xml  classes.dex  diva-beta.apk  lib  META-INF  res  resources.arsc

┌──(sabyasachi㉿DESKTOP-JFB26PG)-[/mnt/c/Users/sabya/3D Objects/diva]
└─$ |
```

Figure 1.4 Inside of an APK.

res/: This folder has device configurations, bitmaps and layouts, icons, and other data that app developers might like to access from their code. It is a short form of resources.

resources.arsc: A file containing binaries of compiled components which might include images, strings, or other data used by an app. It contains all the resources, including library resources which are likely to be used in the app .arsc file stands for Android Resources.

META-INF/: This folder contains the metadata about the Java package carried by the Jar file. It contains files like MANIFEST.MF, CERT.SF, CERT.RSA, etc. It also carries the certificate of the application and files with their hashes.

lib/: As the name suggests, it contains all the third-party libraries for various CPU architectures of Android devices, such as ARM, x86, and MIPS.

```
GNU nano 6.4                    AndroidManifest.xml
```

Figure 1.5 AndroidManifest.xml file.

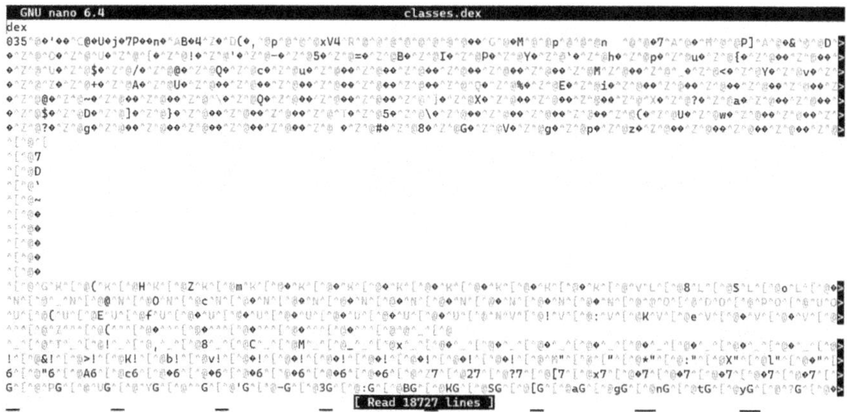

Figure 1.6 classes.dex files.

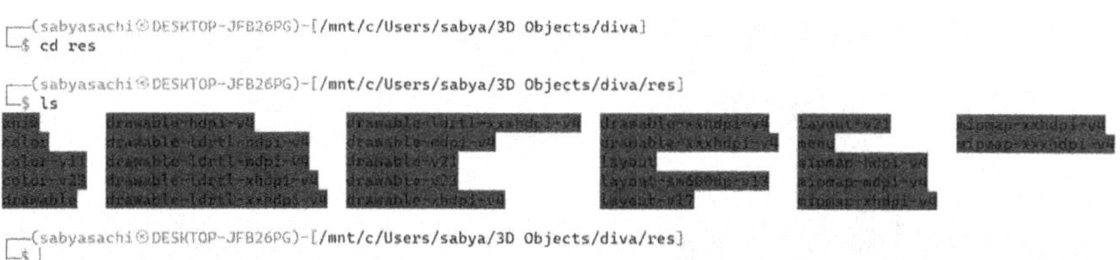

Figure 1.7 res/ folder.

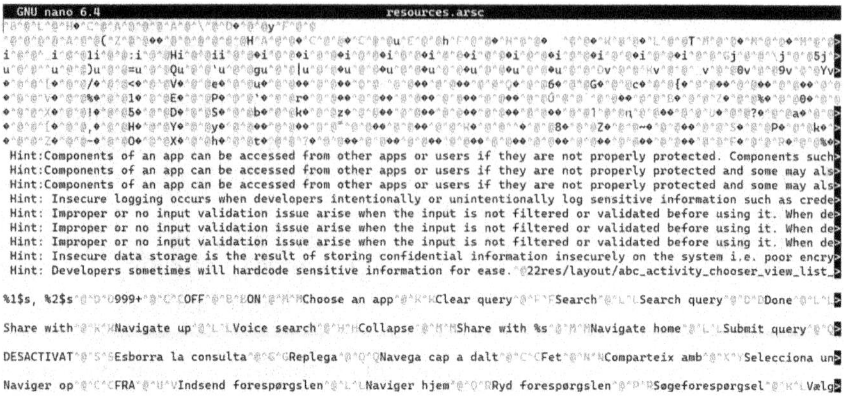

```
GNU nano 6.4                                  resources.arsc
 Hint:Components of an app can be accessed from other apps or users if they are not properly protected. Components such>
 Hint:Components of an app can be accessed from other apps or users if they are not properly protected and some may als>
 Hint:Components of an app can be accessed from other apps or users if they are not properly protected and some may als>
 Hint: Insecure logging occurs when developers intentionally or unintentionally log sensitive information such as crede>
 Hint: Improper or no input validation issue arise when the input is not filtered or validated before using it. When de>
 Hint: Improper or no input validation issue arise when the input is not filtered or validated before using it. When de>
 Hint: Improper or no input validation issue arise when the input is not filtered or validated before using it. When de>
 Hint: Insecure data storage is the result of storing confidential information insecurely on the system i.e. poor encry>
 Hint: Developers sometimes will hardcode sensitive information for ease.^@22res/layout/abc_activity_chooser_view_list_>

%1$s, %2$s^@^D^D999+^@^C^COFF^@^B^BON^@^M^MChoose an app^@^K^KClear query^@^F^FSearch^@^L^LSearch query^@^D^DDone^@^L^L>

Share with^@^K^KNavigate up^@^L^LVoice search^@^H^HCollapse^@^M^MShare with %s^@^M^MNavigate home^@^L^LSubmit query^@^Q>

DESACTIVAT^@^S^SEsborra la consulta^@^G^GReplega^@^Q^QNavega cap a dalt^@^C^CFet^@^N^NComparteix amb^@^X^YSelecciona un>

Naviger op^@^C^CFRA^@^U^VIndsend forespørgslen^@^L^LNaviger hjem^@^Q^RRyd forespørgslen^@^P^RSøgeforespørgsel^@^K^LVælg>
```

Figure 1.8 resources.arsc file.

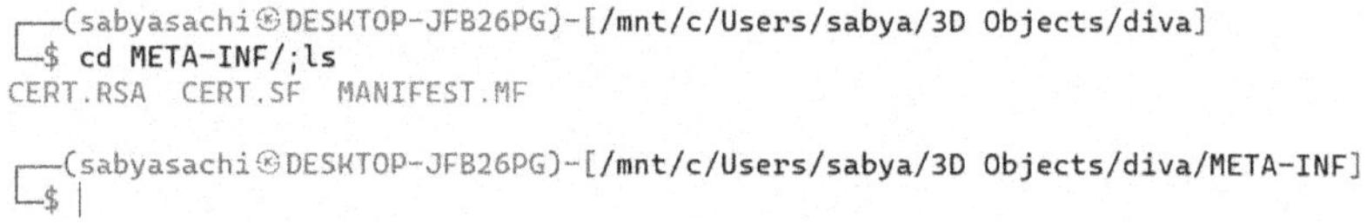

```
┌──(sabyasachi㉿DESKTOP-JFB26PG)-[/mnt/c/Users/sabya/3D Objects/diva]
└─$ cd META-INF/;ls
CERT.RSA  CERT.SF  MANIFEST.MF

┌──(sabyasachi㉿DESKTOP-JFB26PG)-[/mnt/c/Users/sabya/3D Objects/diva/META-INF]
└─$ |
```

Figure 1.9 META-INF/ folder.

```
┌──(sabyasachi㉿DESKTOP-JFB26PG)-[/mnt/c/Users/sabya/3D Objects/diva]
└─$ ls
1.png  2.png  3.png  4.png  AndroidManifest.xml  classes.dex  diva-beta.apk  lib  META-INF  res  resources.arsc

┌──(sabyasachi㉿DESKTOP-JFB26PG)-[/mnt/c/Users/sabya/3D Objects/diva]
└─$ cd lib/;ls
arm64-v8a  armeabi  armeabi-v7a  mips  mips64  x86  x86_64

┌──(sabyasachi㉿DESKTOP-JFB26PG)-[/mnt/c/Users/sabya/3D Objects/diva/lib]
└─$ cd arm64-v8a/

┌──(sabyasachi㉿DESKTOP-JFB26PG)-[/mnt/c/Users/sabya/3D Objects/diva/lib/arm64-v8a]
└─$ ls
libdivajni.so

┌──(sabyasachi㉿DESKTOP-JFB26PG)-[/mnt/c/Users/sabya/3D Objects/diva/lib/arm64-v8a]
└─$ cat libdivajni.so
□
 □□□□□□□□□□□□p□p□p□$$□Q♦td□R♦td□♦

□`□□□□♦□♦□□□□□♦□□♦□L□
                      □□♦□□□♦□□□C□□□♦□□□♦□□□K□♦□l♦□□♦□□□__cxa_finalize__cxa_atexitJava_jakhar_aseem_diva_DivaJni_accessst
rncmpJava_jakhar_aseem_diva_DivaJni_initiateLaunchSequencestrcpyJNI_OnLoadlibstdc++.solibm.solibc.solibdl.so_edata__bss_
□□□`□♦□□□□♦□□□♦□□□__bss_end____end___endlibdivajni.so♦
♦□□□
```

Figure 1.10 lib/ folder.

1.5 APPLICATION SIGNING

While researching Android development and pentesting, any developer, pentester, or simply an enthusiast frequently encounters the subject of signing applications. While this is a simple concept in and of itself, the definitions and theories found online can make it appear to be quite challenging. Let me now describe the signing of an application to you while keeping the newcomers in mind. Before an app can be installed on Android, it must be digitally signed with a certificate. Android does not require the certificate to be signed by a certificate authority in order to use it to identify the author of an app. Self-signed certificates are commonly used by Android apps. The private key for the certificate is kept by the app developer. The foundation of Android app security is the use of a produced certificate and digital "key" that offers a distinctive, encrypted, and largely impenetrable signature. This demonstrates that the app originated from you and not from a questionable source. This is accomplished using an Android keystore. The keystore is a straightforward file that contains a sizable chunk of encrypted data. This is typically the first issue that developers run into because this file could be located anywhere on your machine. You should be aware of the differences between release and debug keystores. In addition, keystore files are secured by two passwords, one for the main keystore file and the other for each keystore/alias pair contained within the file. The majority of developers use the same password for both, even though these passwords should ideally be distinct. Apps can be signed in release or debug modes. When developing your program, you sign it in a debug mode, and when you're ready to release it, you sign it in a release mode. In order to sign programs in a debug mode, the Android SDK (Software Development Kit) generates a certificate. You must create your own certificate in order to sign applications in a release mode.

1.6 APPLICATION SANDBOXING MODEL

The Linux user-based protection is utilized by the Android platform to identify and isolate program resources. As a result, apps are kept apart from one another, and the system is shielded against dangerous apps. Android does this by giving each Android application a distinct user ID (UID) and running each one in a separate process. Android creates a kernel-level Application Sandbox using the UID. Through common Linux features like app-assigned user and group IDs, the kernel ensures security between programs and the system at the process level. By default, apps have limited access to the OS and are unable to communicate with one another. App A cannot perform malicious actions because it lacks the necessary default user capabilities, such as reading program B's data or dialing a phone number without authorization. The sandbox is easy to use, auditable, and built on UNIX-style user separation of processes and file rights that have been around for decades. Since the Application Sandbox is a part of the kernel, both native code and OS apps are protected by this security architecture. The Application Sandbox is where all software that runs on top of the kernel, including OS libraries, application frameworks, application runtimes, and all applications, runs. On some systems, programmers are limited to a particular language, API set, or development framework. For the purposes of security on Android, there are no limitations on how an application may be created; hence, native code is just as sandboxed as interpreted code.

Chapter 2

Preparing the pentesting lab

Before jumping into the building lab, the foremost important thing we should be aware of is that there are two types of testing:

Static Analysis: In this method, the source code of the application is analyzed, and based upon that, the VAPT is done.
Dynamic Analysis: In this approach, the app is being run on a device, and then the VAPT is done.

Since the majority of readers use Windows, I will only demonstrate the tests on a Windows system. Since most of the tests will be performed on virtual machines, Mac users can follow along as well, but I'll undoubtedly utilize a Mac in my upcoming book on iOS pentesting.

The softwares we will be installing are as follows:

Virtualization: Oracle VirtualBox, VMWare Workstation (Choose any 1, the installation has been shown for both the softwares keeping the choices of readers in mind)
Attacker's OS: Kali Linux VM Images
Tools and Softwares: Android Emulator, Android Studio, adb, Jadx & Jadx GUI, keytool, apktool, Java, Dex2Jar, and Burp Suite as of now. If more tools are needed, we'll install them on the go.
Vulnerable Apps: DIVA, InsecureBankv2, and Sieve

Before moving ahead to the installation part, let's just quickly first understand the importance of virtualization with regard to mobile penetration testing.

In order to create a safe and adaptable testing environment, virtualization is essential for mobile penetration testing. Virtualization is crucial for mobile penetration testing for the following main reasons:

1. Security and isolation: Using virtualization, you can build secure, isolated virtual machines (VMs) just for testing. By simulating various mobile devices and operating systems, these VMs let you do penetration testing without jeopardizing the security of real devices. The risk

DOI: 10.1201/9781003450474-2

of potential malware or vulnerabilities found during testing spreading to other systems is reduced by isolating the testing environment, which keeps them restricted within the virtual environment.

2. Device diversity refers to the range of models, operating systems, and settings available for mobile devices. With the help of virtualization, you may imitate and test on several virtual devices at once, covering a wide range of device kinds and settings. This variety ensures that the penetration testing is thorough and accounts for a variety of scenarios.
3. Rapid deployment and scalability: Virtualization makes it possible to quickly deploy VMs, giving testers the ability to quickly set up and customize various mobile environments. Additionally, it offers scalability, allowing you to easily add more virtual machines as needed without having to buy and maintain several actual devices. Due to the time and resource savings provided by this flexibility, penetration testers are better able to concentrate on the testing itself.

As a result of its ability to create a secure, scalable, and varied testing environment, virtualization is crucial to mobile penetration testing. It makes it possible for testers to successfully cooperate, isolate vulnerabilities, and test on a variety of platforms and settings. Penetration testers can improve the effectiveness and efficiency of their mobile security tests by utilizing virtualization.

Let's move on to installing the aforementioned prerequisites right away, and for ease of access, the software links will also be provided.

2.1 INSTALLING ORACLE VIRTUALBOX

Link: https://www.virtualbox.org/wiki/Downloads

Step #1: Go to Google, search for "install oracle virtualbox," and open the link

Step #2: Click on the Downloads button, select Windows hosts, and also download the extension pack.

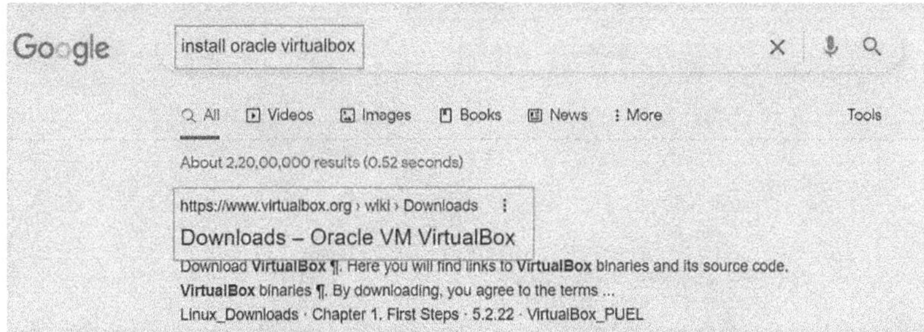

Figure 2.1 Downloading Oracle Virtualbox using Google.

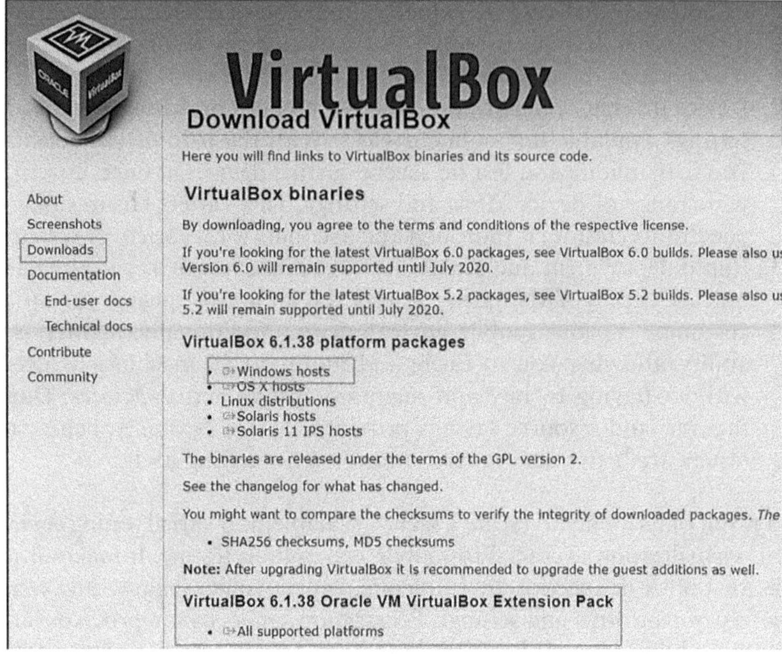

Figure 2.2 Downloading the VBox software and extension pack.

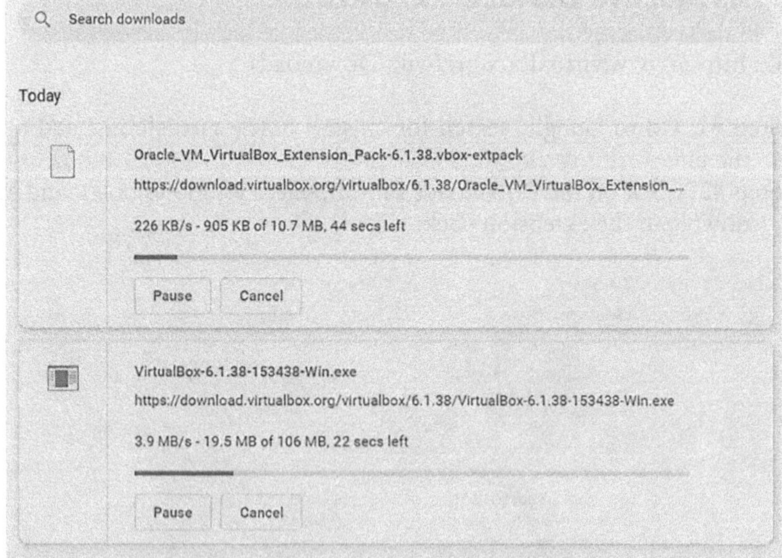

Figure 2.3 Download page of the browser.

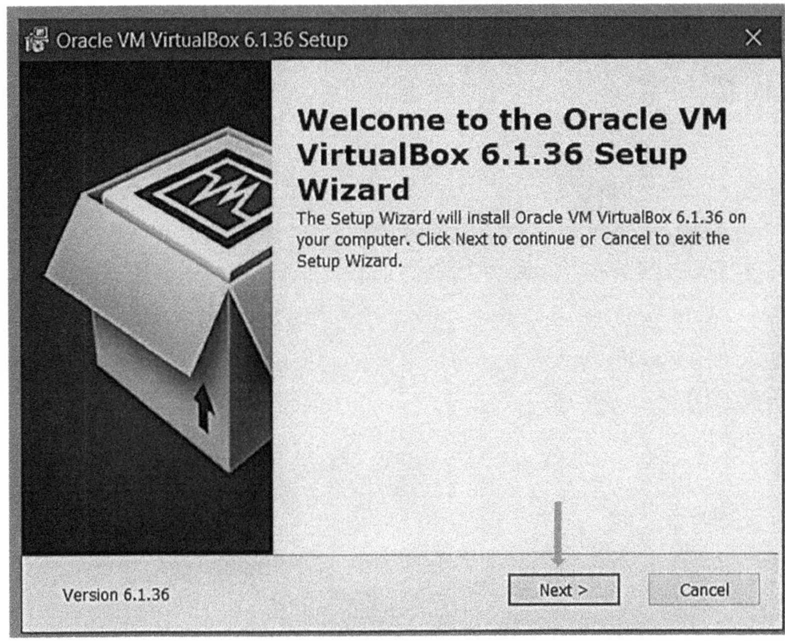

Figure 2.4 Installation window of the VBox.

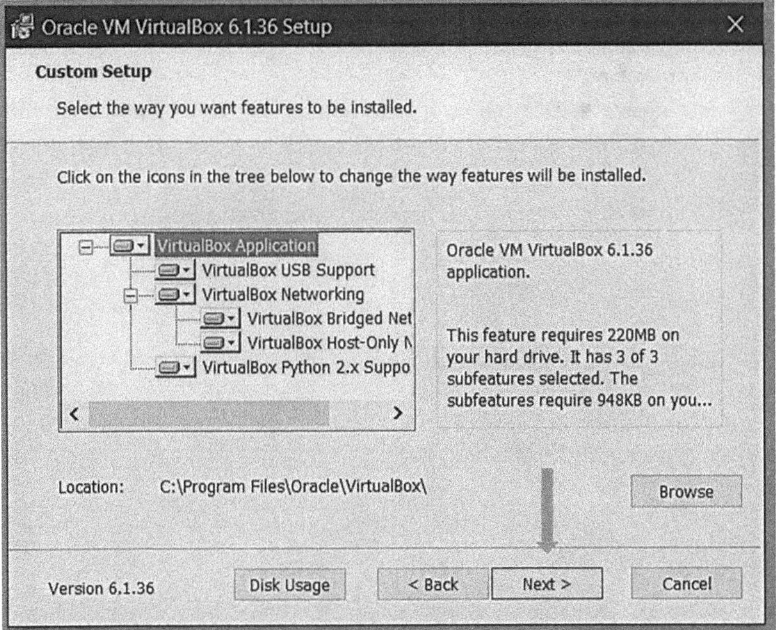

Figure 2.5 Setup window during VBox installation.

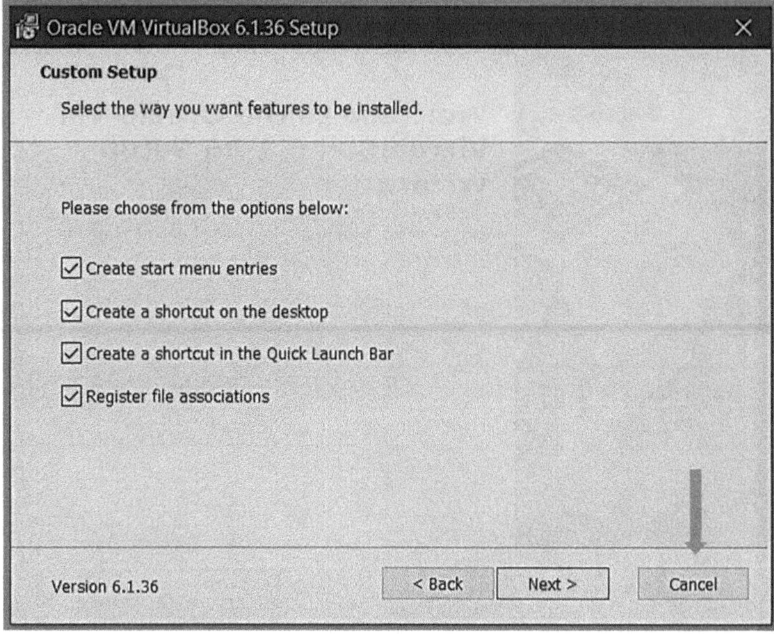

Figure 2.6 Feature selection window during VBox installation.

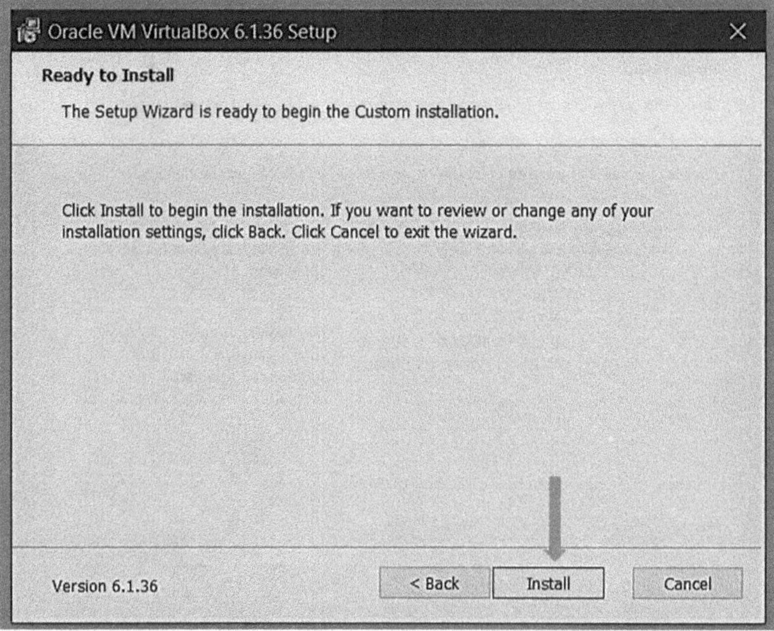

Figure 2.7 Installation prompt.

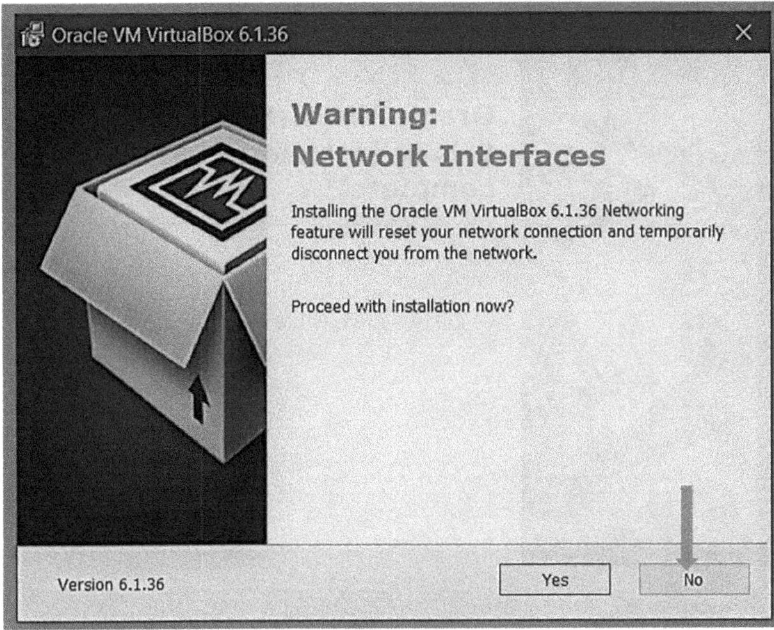

Figure 2.8 Network Adapter Installation Permission.

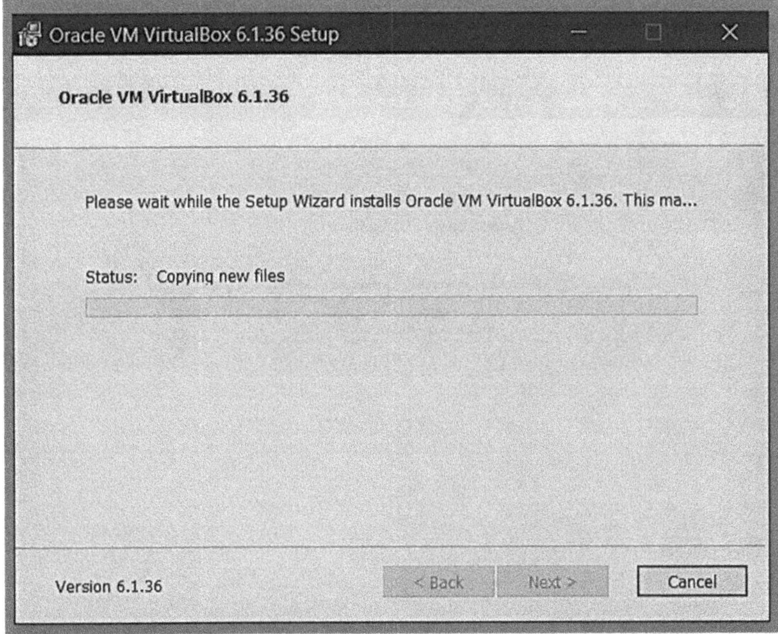

Figure 2.9 Installation and setting up of files.

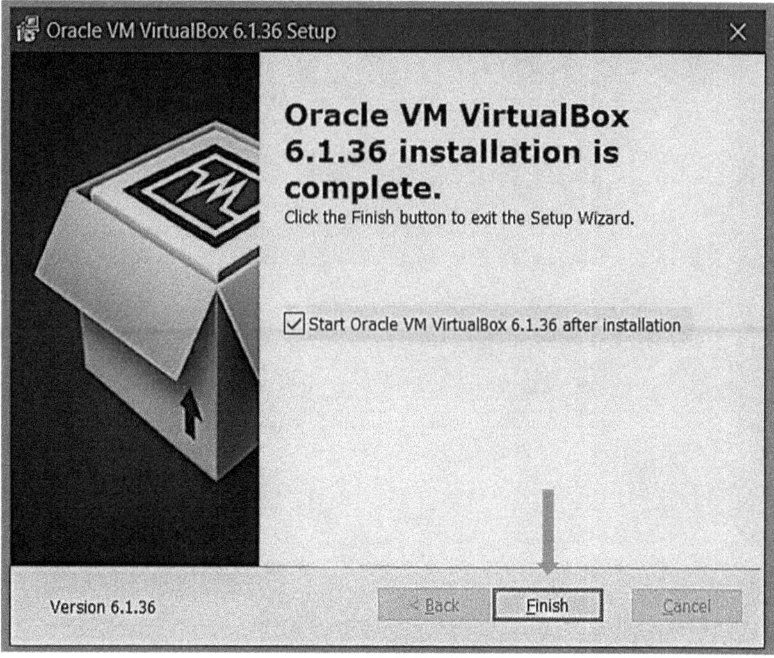

Figure 2.10 Installation completed confirmation prompt.

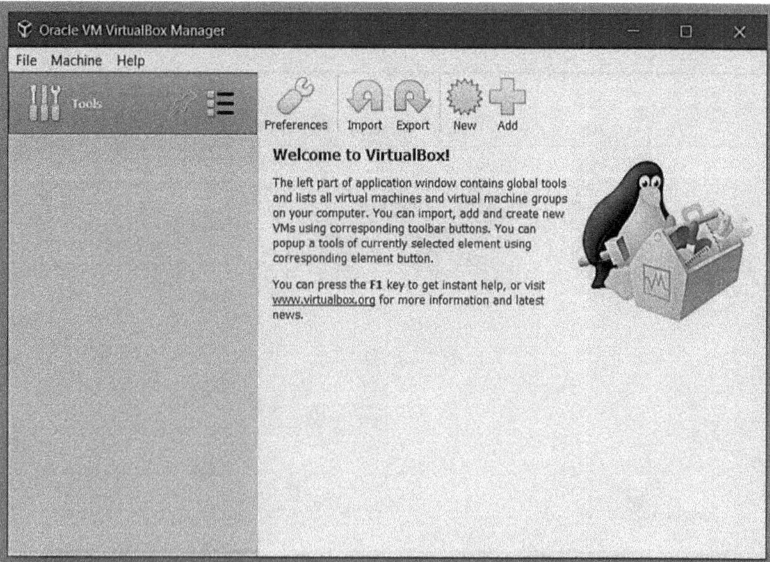

Figure 2.11 Oracle Virtualbox window.

Your download will start

Step #3: Open the executable file and install it. Click Next to begin installation.

Step #4: Click Next, next couple of times, and finally click Finish

Step #5: Double click on the icon in the Desktop, and this is how VBox looks

2.2 INSTALLING VMWARE

Link: https://customerconnect.vmware.com/en/downloads/details?downloadGroup=WKST-PLAYER-1624&productId=1039&rPId=91446

Step #1: Go to Google, search install "vmware workstation player," and click on the link

Step #2: Click on the Download button, and it will redirect you to a page

Step #3: Click on the Go to Downloads button, and a page will open

Step #4: Click the download options for Windows

Your download will start.

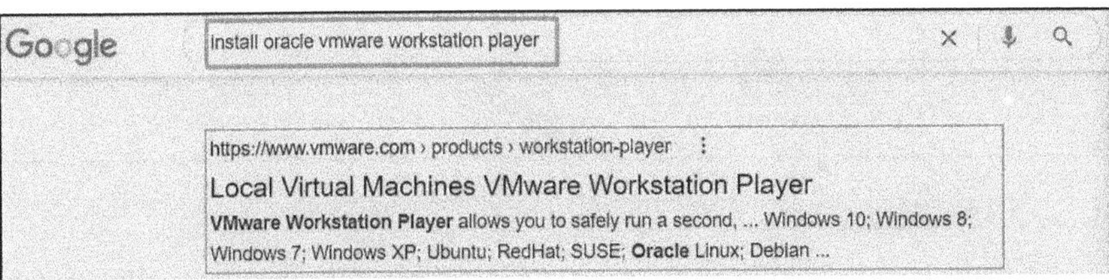

Figure 2.12 Downloading VMWare workstation player.

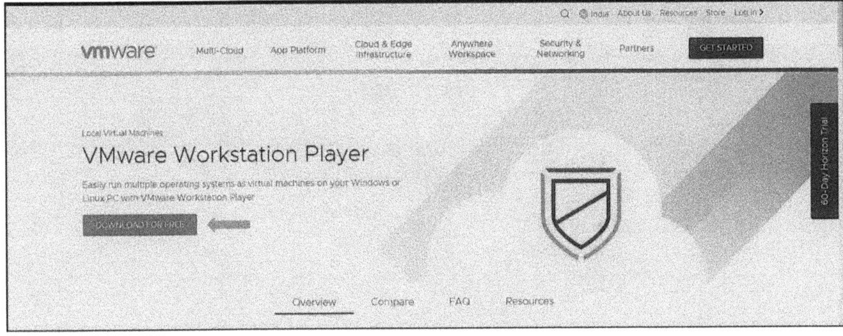

Figure 2.13 Downloading VMWare workstation player.

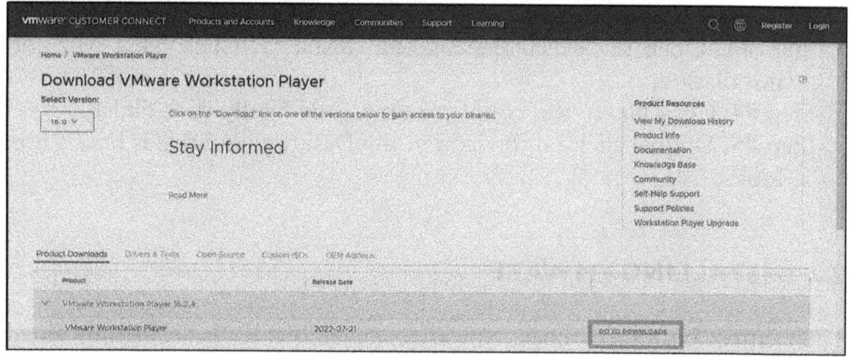

Figure 2.14 Customer support page of Vmware player.

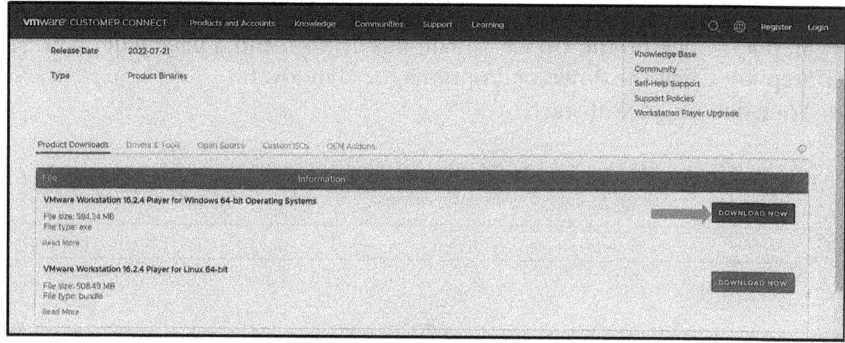

Figure 2.15 Executable download page.

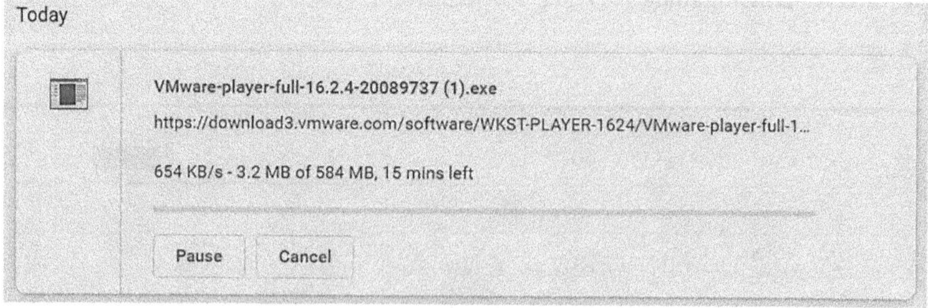

Figure 2.16 Download page.

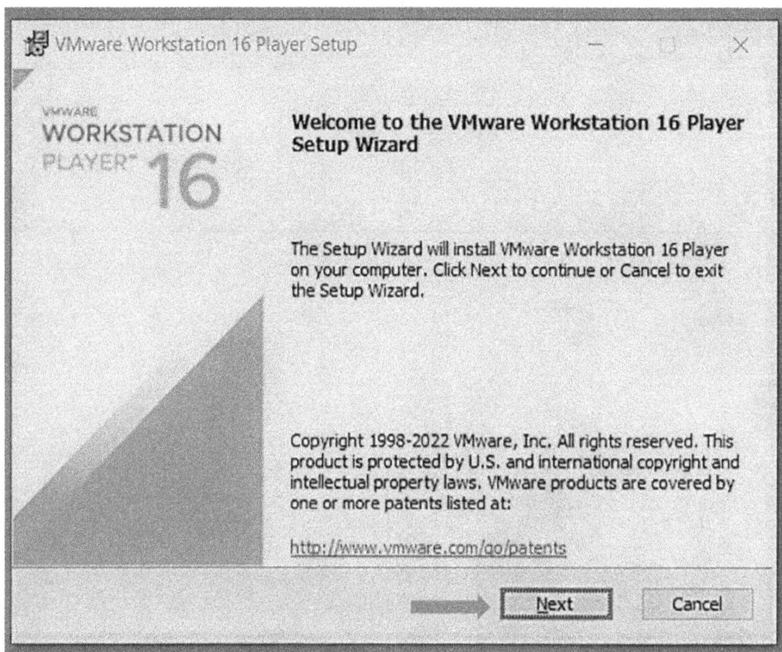

Figure 2.17 Installation window.

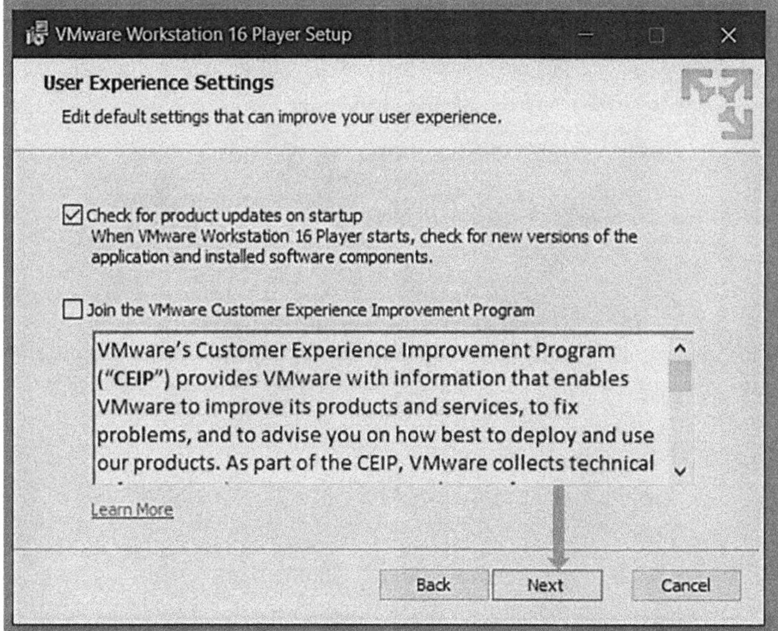

Figure 2.18 User experience settings.

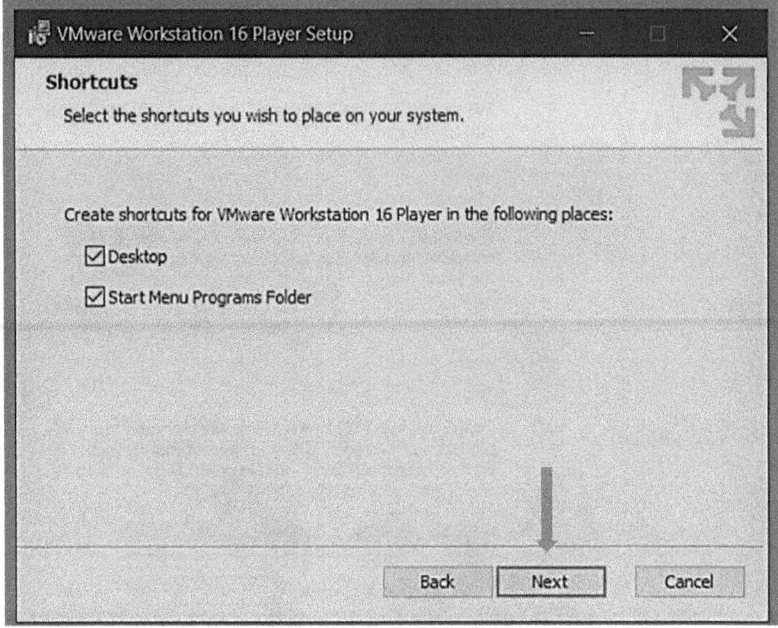

Figure 2.19 Desktop shortcuts.

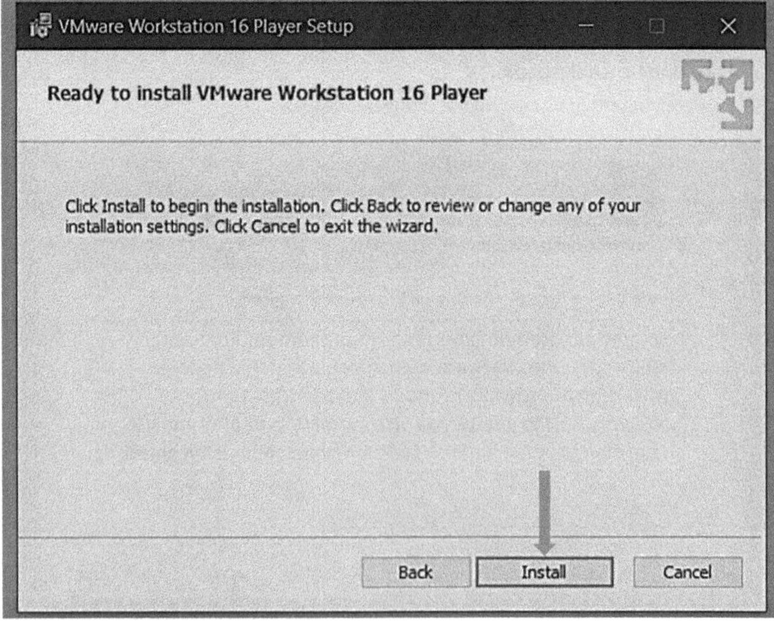

Figure 2.20 Installation beginning.

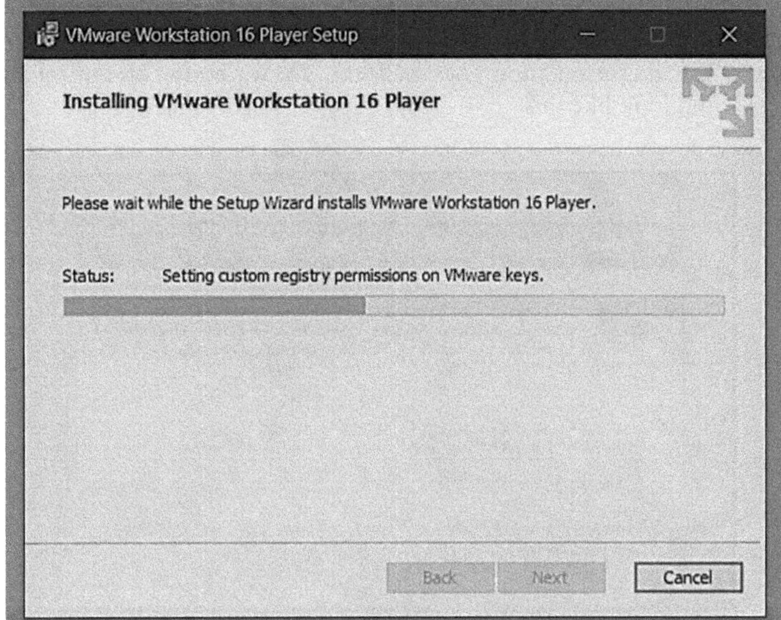

Figure 2.21 During the installation.

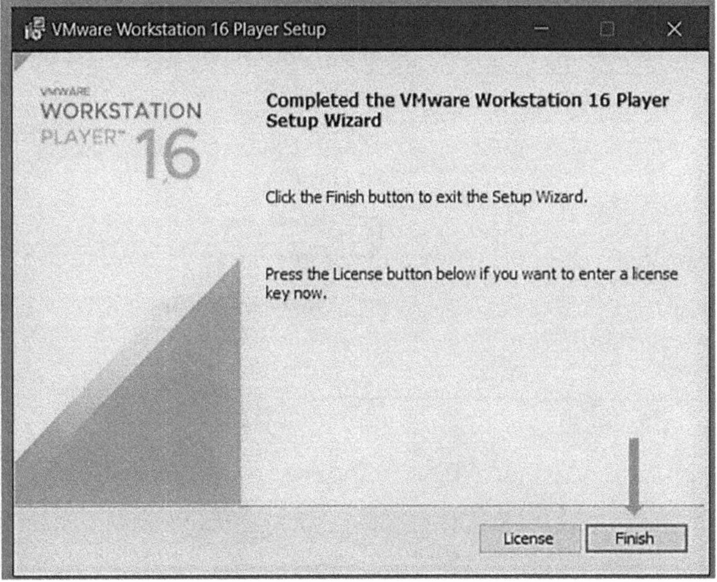

Figure 2.22 Installation confirmed window.

Step #5: Once downloaded, open the executable to install it. Click Next.
Step #6: Keep clicking Next, Next a couple of times and Finish at last.
Step#7: If the installation goes well, the VMWare after the installation should look like this

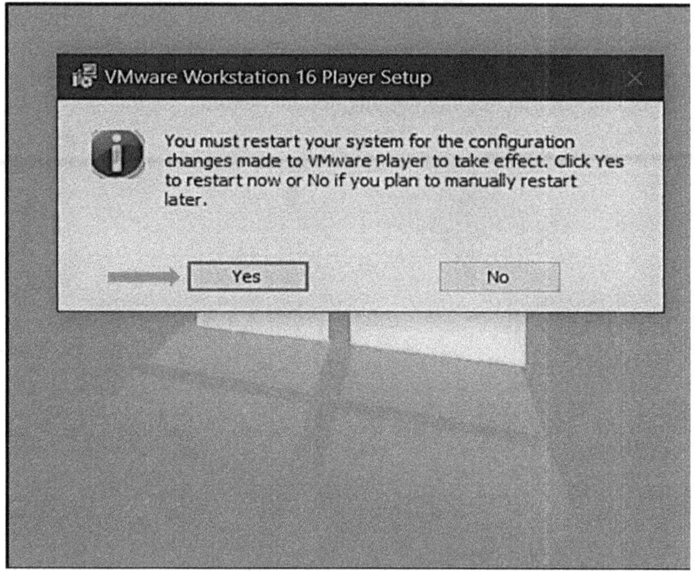

Figure 2.23 Reboot notice for updating drivers and adapters.

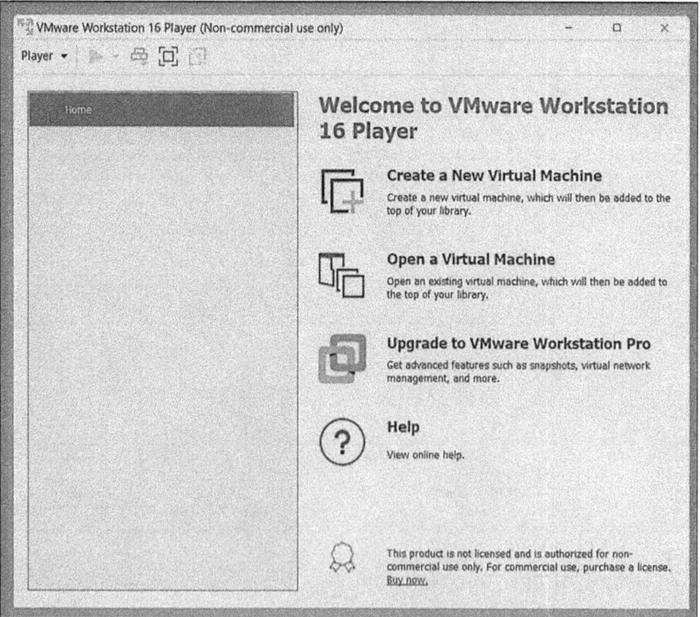

Figure 2.24 VMWare workstation player window.

2.3 SETTING UP OF KALI LINUX

Link: https://www.kali.org/get-kali/#kali-virtual-machines

For Oracle VirtualBox:

Step #1: Go to Google and search for Kali. Click on the link to go to the official site.
Step #2: In the official site of Kali, click on the Download button.
Step #3: Select the virtual platform option.
Step #4: Choose the virtual box option.
Step #5: Your Kali image will start to download.
Step #6: Open the explorer and extract the zip file.
Step #7: Your zip file will start extracting.
Step #8: Once extracted, open Virtualbox and click on the Add button.
Step #9: Select the VBox file from the extracted folder and click Open.
Step #10: After it is being loaded, click on the Start button to run the VM.
Step #11: Wait for the OS to load and enter username/password as kali/kali.

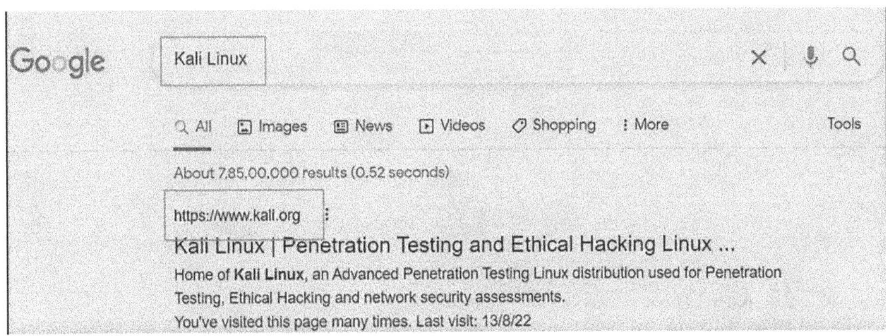

Figure 2.25 Downloading Kali Linux.

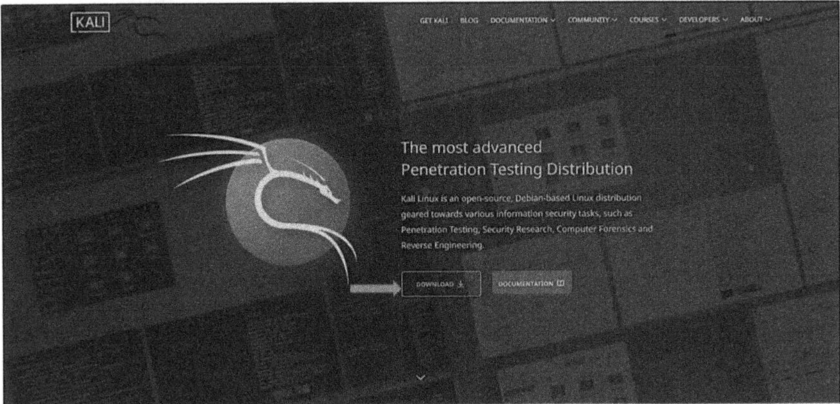

Figure 2.26 Official Kali website.

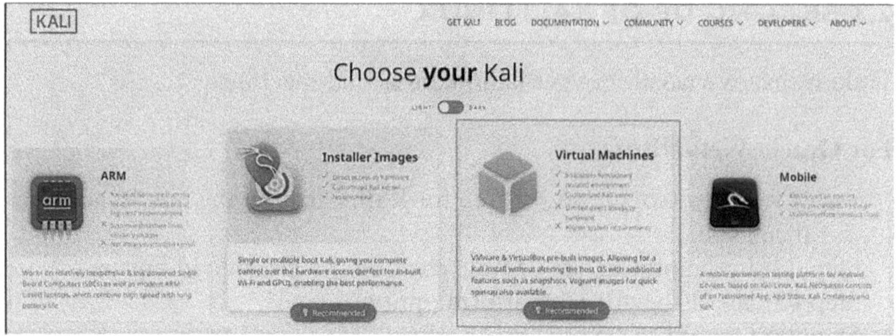

Figure 2.27 Kali images for different platforms

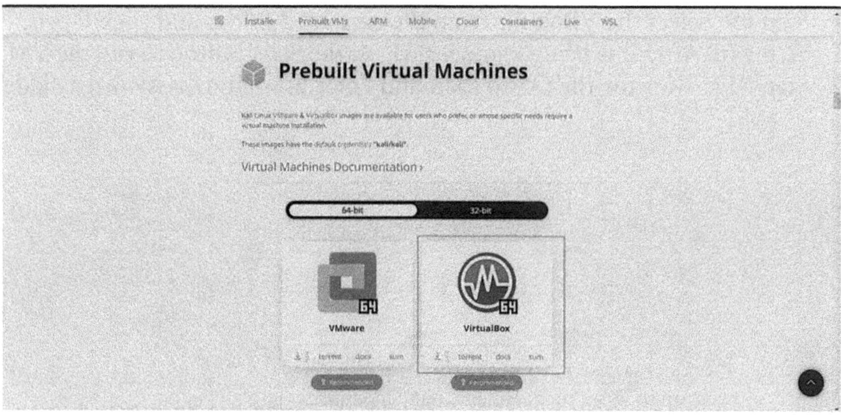

Figure 2.28 Kali Linux Virtualbox platform.

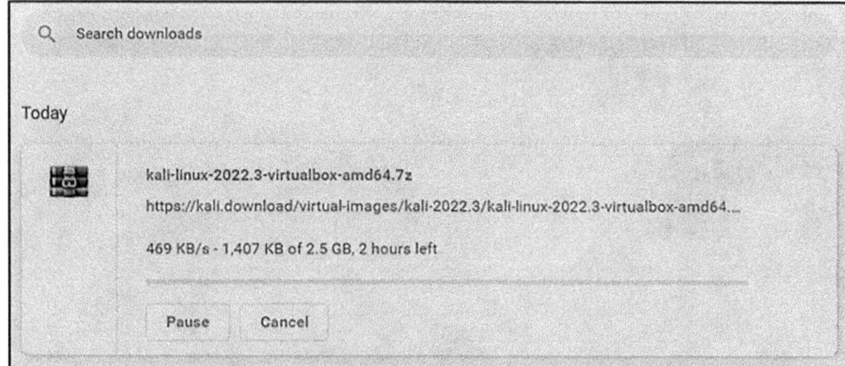

Figure 2.29 Downloading the Kali VBox image.

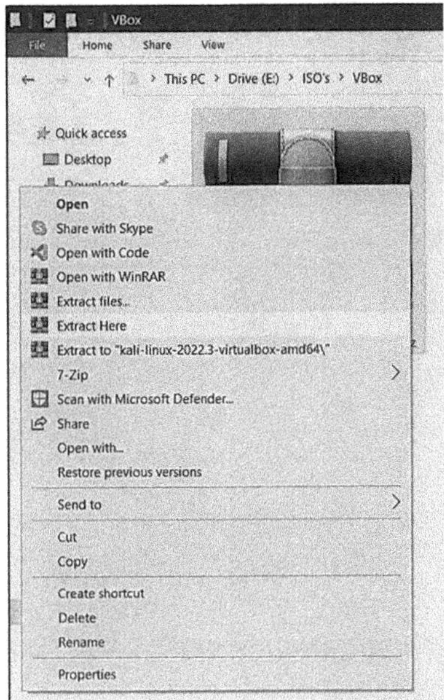

Figure 2.30 The extract option of the zip file.

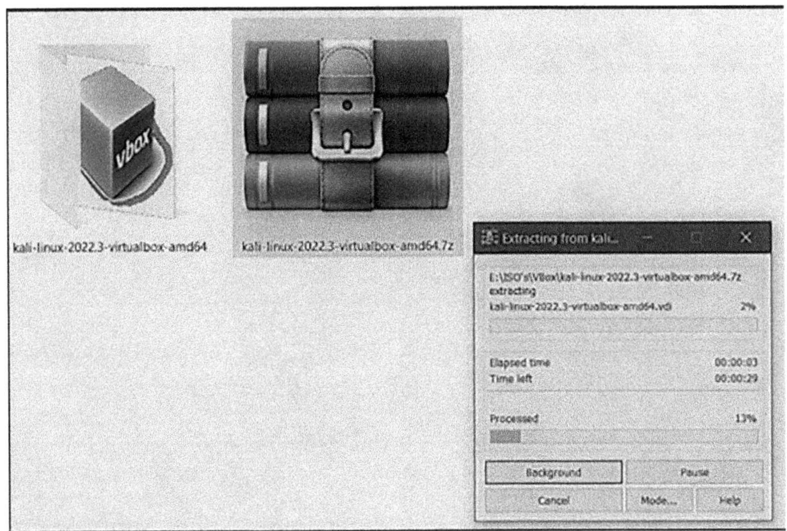

Figure 2.31 Zip file is being extracted.

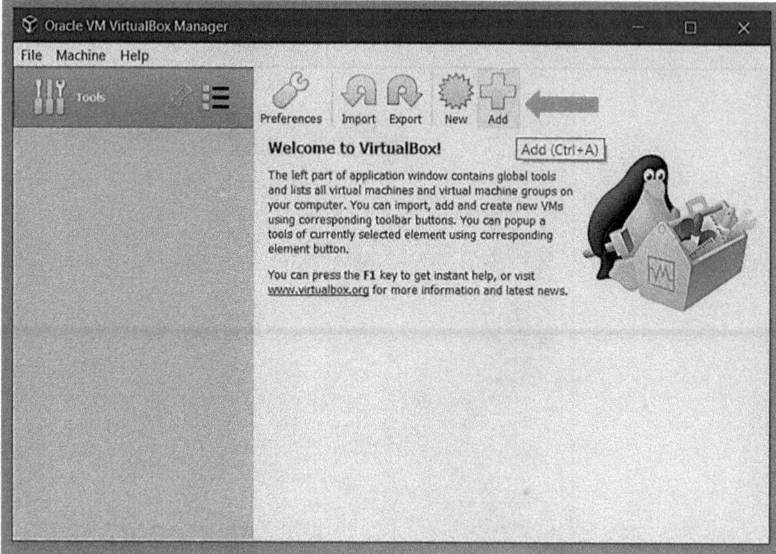

Figure 2.32 Loading Kali into VBox.

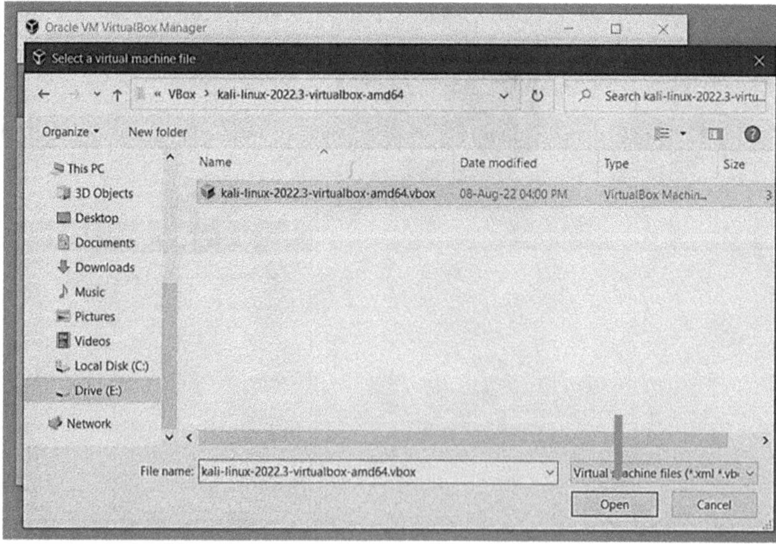

Figure 2.33 Selecting the VBox image.

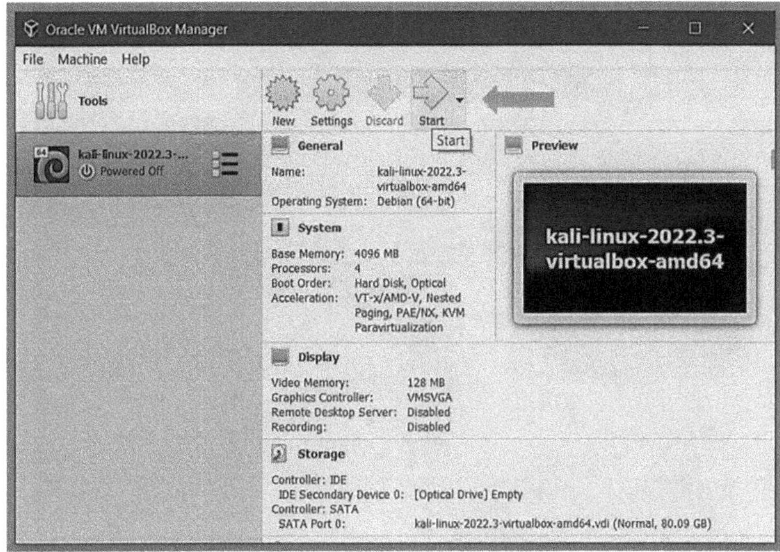

Figure 2.34 Kali loaded on VBox.

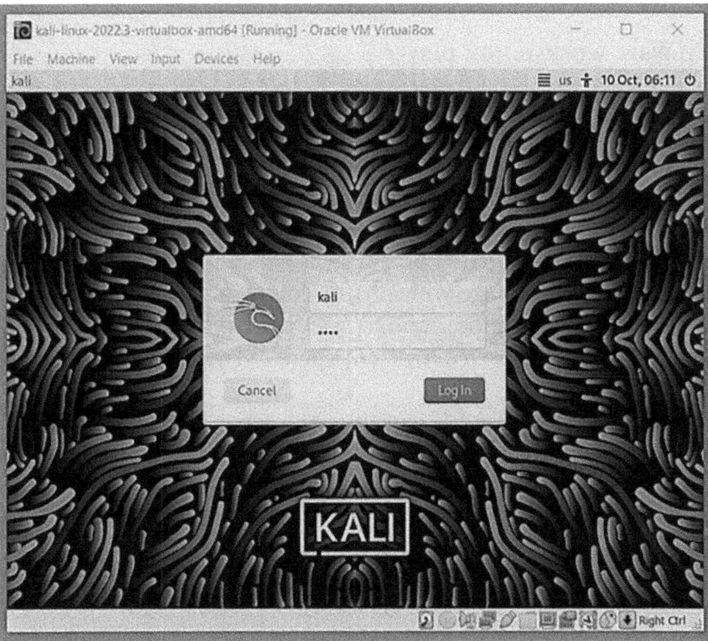

Figure 2.35 Kali login screen.

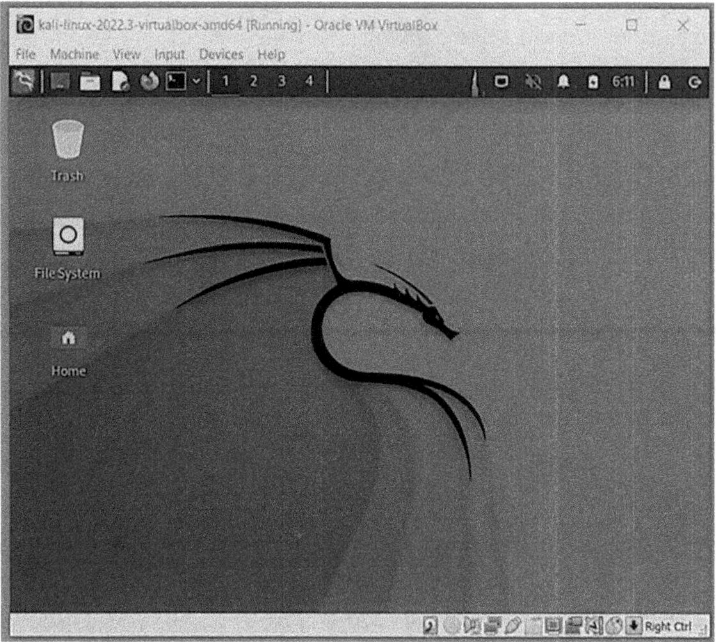

Figure 2.36 The Kali OS.

For VMWare workstation player:

Step #1: Go to Google and search for Kali. Click on the link to go to the official site.
Step #2: In the official site of Kali, click on the Download button.
Step #3: Select the virtual platform option.
Step #4: Choose the virtual box option.
Step #5: Your Kali image will start to download.
Step #6: Open the explorer and extract the zip file.
Step #8: After it has been extracted, open VMware and click on the Open virtual machine.
Step #9: Select the .vmx file from the extracted folder and click Open.
Step #10 (optional): After it has been added, click on Edit virtual machine settings.
Step #11 (Optional): Increasing RAM and Cores of the VM since I am having many.

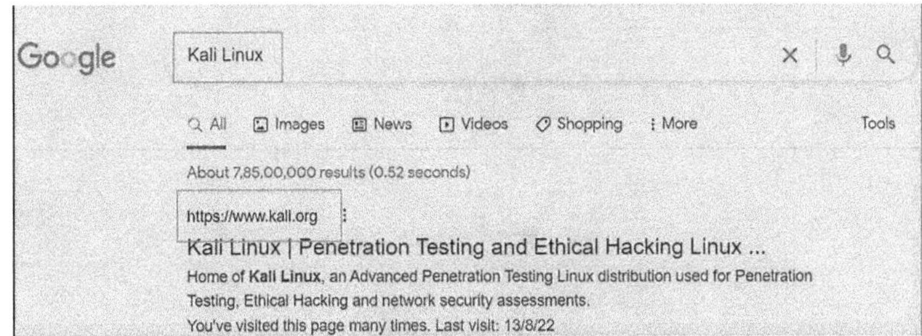

Figure 2.37 Downloading Kali Linux.

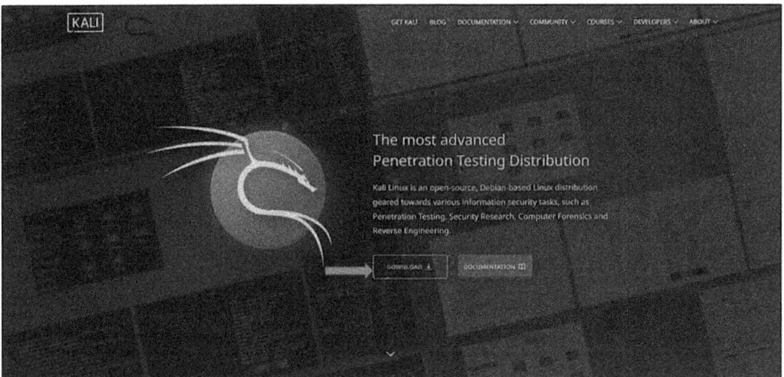

Figure 2.38 Official Kali website.

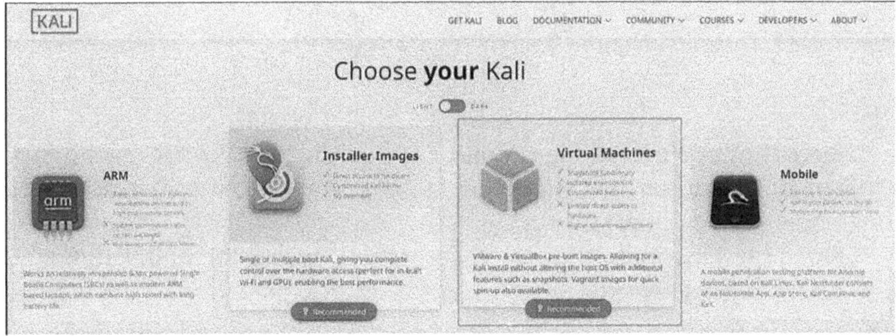

Figure 2.39 Kali images for different platforms.

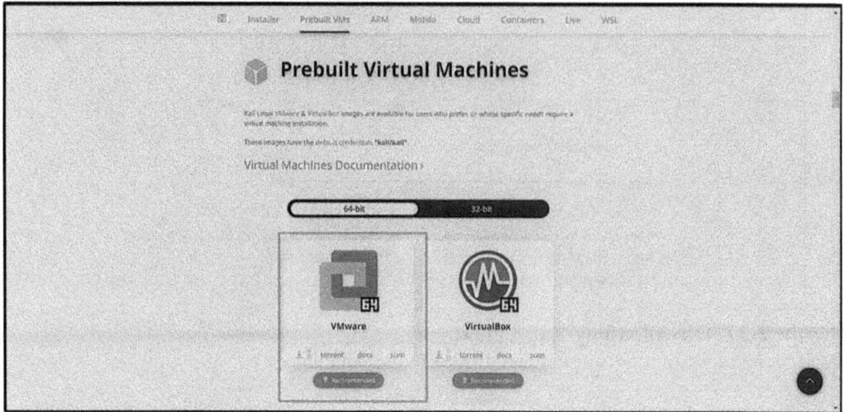

Figure 2.40 Kali Linux VMware platform.

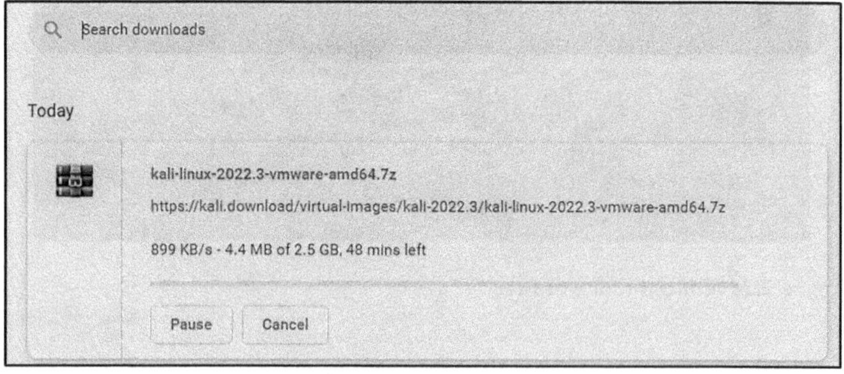

Figure 2.41 Downloading the Kali VMWare image.

Step #12: Once everything is done, click on the Play Virtual Machine.
Step #13: Wait for the OS to load, and if you get this warning, then click on the second option.
Step #14: Once all is done, the machine will boot and kali/kali is username/password.

2.4 INSTALLING THE TOOLS

Open the Kali instance on any of your favorite virtualization software and open the terminal. Now, we are going to install the tools required for our penetration testing purposes.

Extracting from kali...

E:\ISO's\Vmware\kali-linux-2021.3-vmware-amd64.7z
extracting
Kali-Linux-2021.3-vmware-amd64-s001.vmdk 0%

Elapsed time 00:00:10
Time left 00:08:24

Processed 2%

Background | Pause
Cancel | Mode... | Help

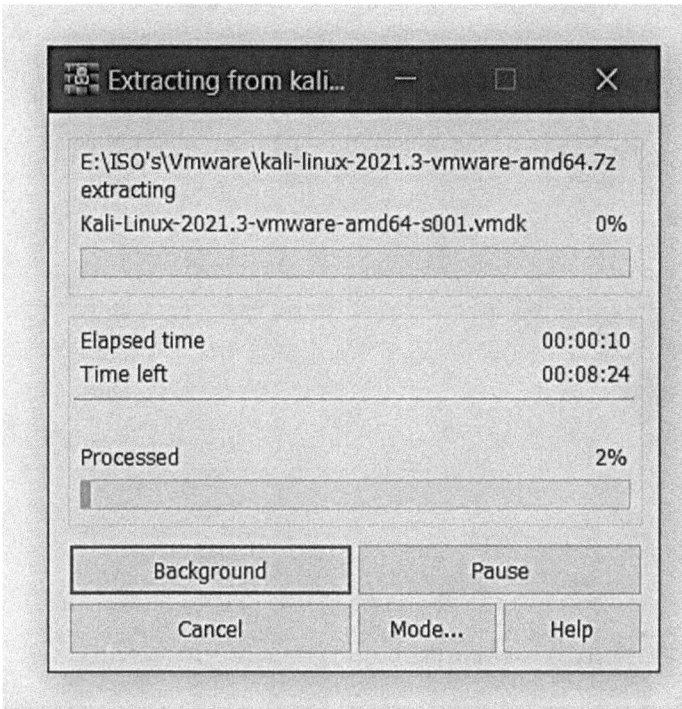

Figure 2.42 The extract option of the zip file.

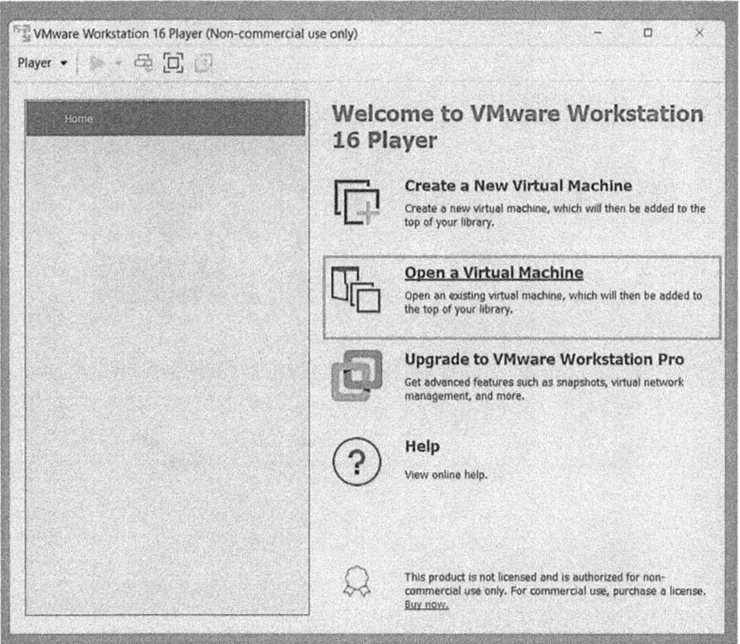

Figure 2.43 Adding the kali iso to vmware.

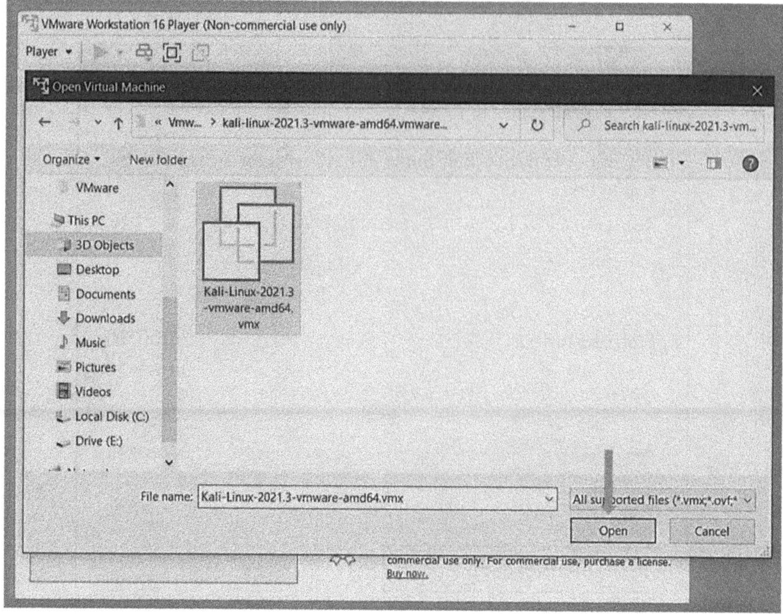

Figure 2.44 Selecting the vmx file.

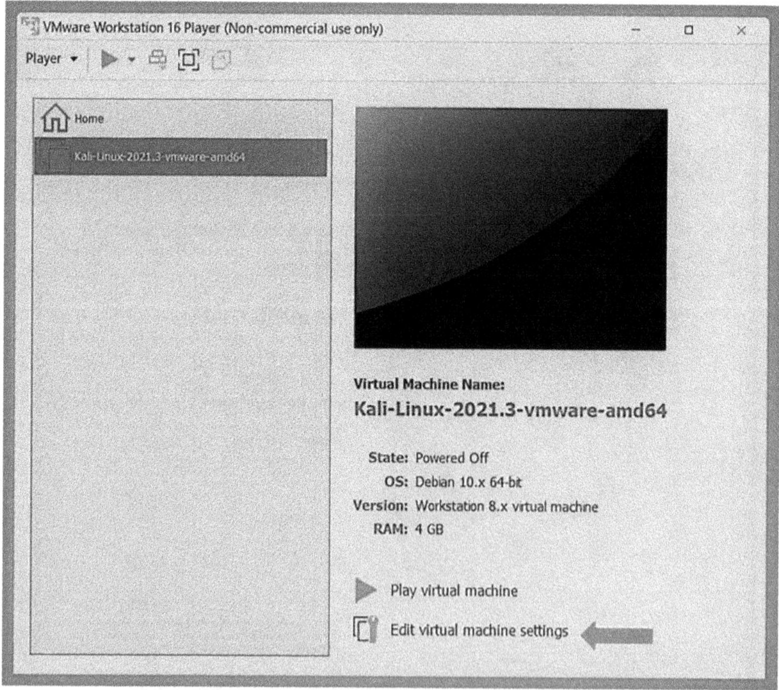

Figure 2.45 Kali Linux loaded in vmware.

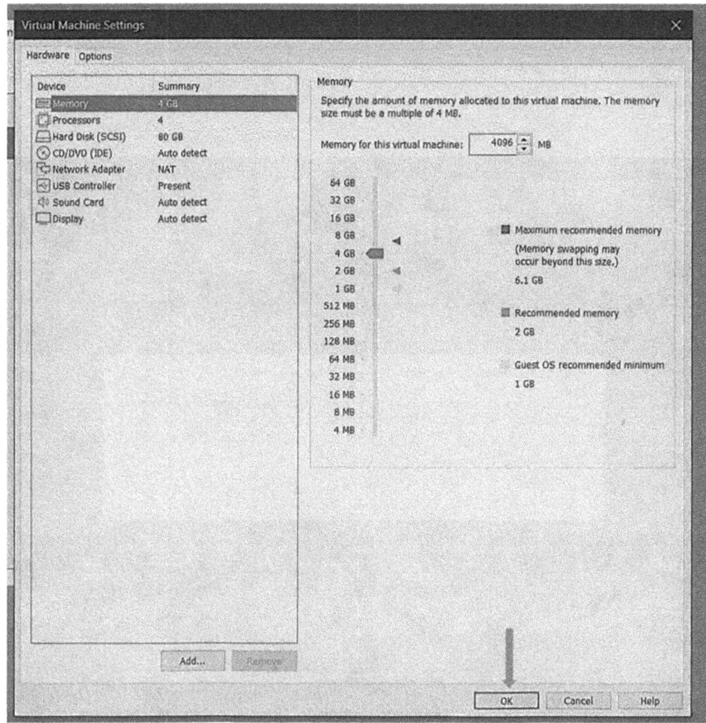

Figure 2.46 Increasing memory size of the VM.

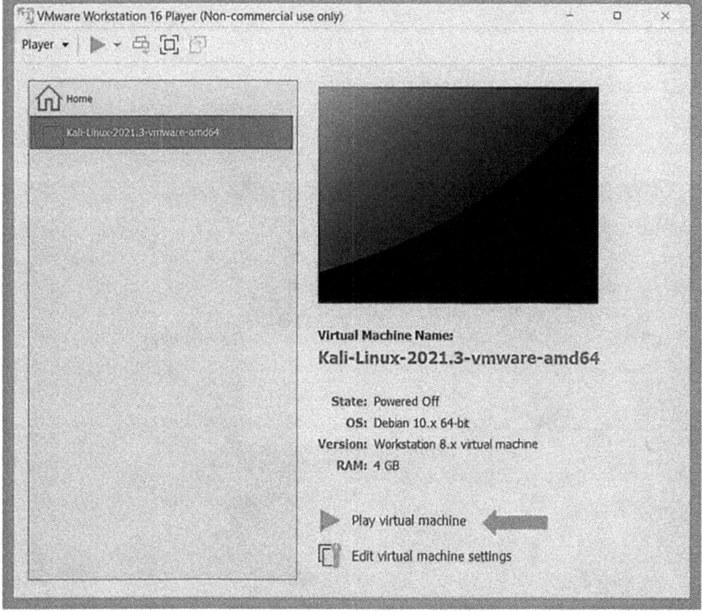

Figure 2.47 Playing the VM.

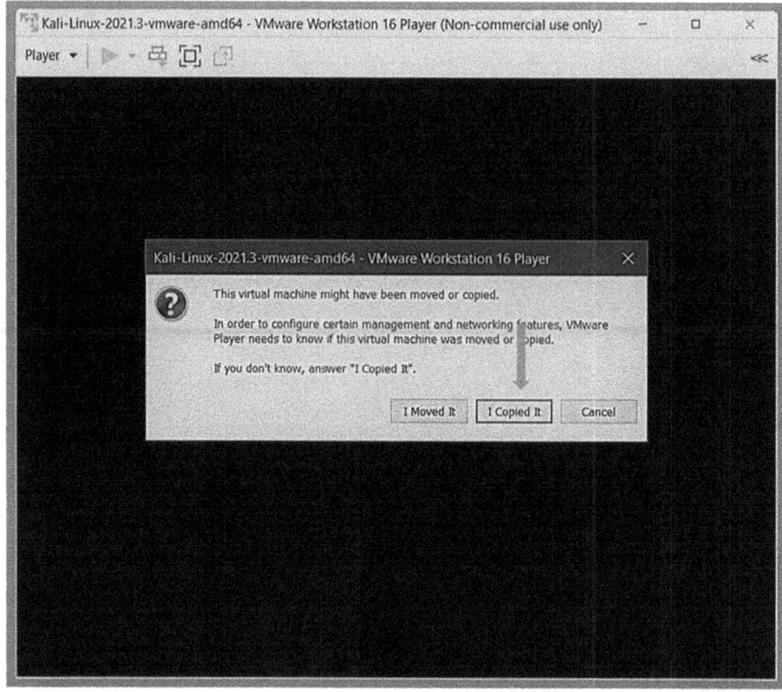

Figure 2.48 HERE> Management configuration prompt.

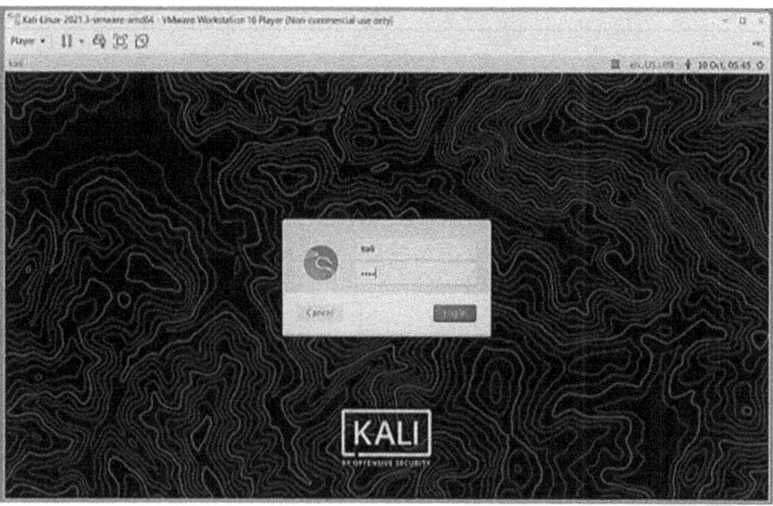

Figure 2.49 Machine is booting up.

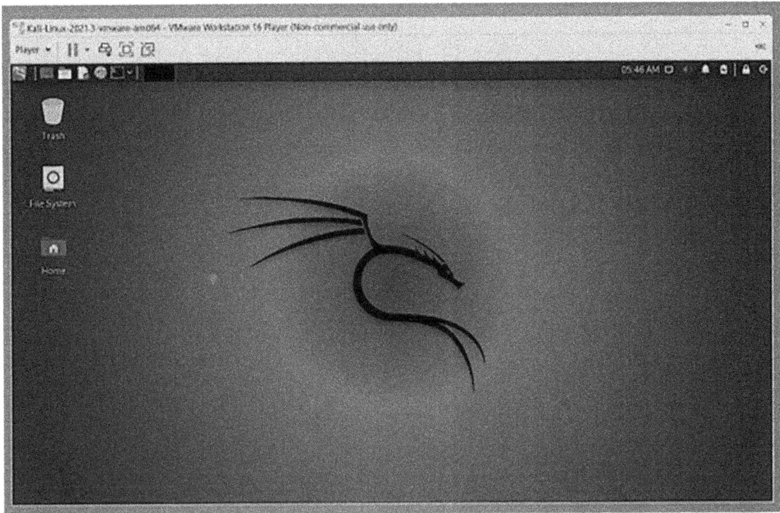

Figure 2.50 The Kali OS.

Note: At first all the tools might not install, try executing 'sudo apt update' without quotes and then install the tools.

2.4.1 Installing JADX-GUI

Step #1: Open terminal and type: *sudo apt install jadx*.
Step #2: Once done, both jadx CLI and GUI versions have been installed.

2.4.2 Installing apktool

Step #1: In the terminal type: *sudo apt install apktool*
Step #2: Bring the pointer over <No> using Tab and select it using the Enter key.
Step #3: Bring the pointer of <OK> using Tab and hit Enter.
Step #4: If everything goes right, then the installation begins.
Step #5: After installation is complete, you can run apktool by typing *apktool -h*.

2.4.3 Installing Java

For Kali:

Step #1: Type *sudo apt install openjdk-11-jre-headless*.
Step #2: Once installation is done, type *java – version* to verify the installation.

```
(kali㉿kali)-[~]
$ sudo apt install jadx
Reading package lists... Done
Building dependency tree... Done
Reading state information... Done
The following NEW packages will be installed:
  jadx
0 upgraded, 1 newly installed, 0 to remove and 1675 not upgraded.
Need to get 27.0 MB of archives.
After this operation, 30.1 MB of additional disk space will be us
ed.
Get:1 http://kali.download/kali kali-rolling/main amd64 jadx all
1.4.4-0kali1 [27.0 MB]
44% [1 jadx 14.9 MB/27.0 MB 55%]
```

Figure 2.51 Installing Jadx.

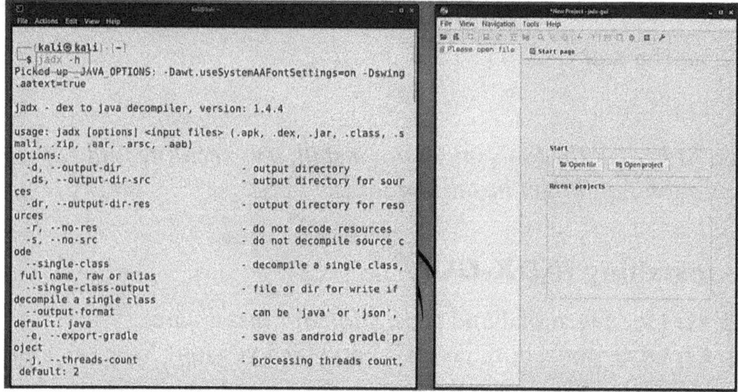

Figure 2.52 Jadx CLI on the left and Jadx-GUI ion the right.

```
(kali㉿kali)-[~]
$ sudo apt install apktool -y
Reading package lists... Done
Building dependency tree... Done
Reading state information... Done
The following additional packages will be installed:
  aapt android-framework-res android-libaapt android-libandroidfw
  android-libbacktrace android-libbase android-libcutils android-liblog
  android-libutils android-libziparchive gcc-12-base junit libantlr-java
  libantlr3-runtime-java libaopalliance-java libapache-pom-java
  libatinject-jsr330-api-java libc-bin libc-dev-bin libc-l10n libc6 libc6-dev
  libc6-i386 libcdi-api-java libcommons-cli-java libcommons-compress-java
  libcommons-io-java libcommons-lang3-java libcommons-parent-java
  libcommons-text-java libgeronimo-annotation-1.3-spec-java
  libgeronimo-interceptor-3.0-spec-java libguava-java libguice-java libjansi-java
  libjsr305-java libmaven-archiver-java libmaven-file-management-java
  libmaven-jar-plugin-java libmaven-parent-java libmaven-resolver-java
  libmaven-shared-io-java libmaven-shared-utils-java libmaven3-core-java
  libplexus-archiver-java libplexus-cipher-java libplexus-classworlds-java
  libplexus-component-annotations-java libplexus-interpolation-java
  libplexus-io-java libplexus-sec-dispatcher-java libplexus-utils2-java
  libprotobuf-lite23 libsisu-inject-java libsisu-plexus-java libslf4j-java
  libsmali-java libsnappy-java libsnappy-jni libstdc++6 libstringtemplate-java
  libwagon-provider-api-java libxmlunit-java libxpp3-java libxz-java
```

Figure 2.53 Installing apktool.

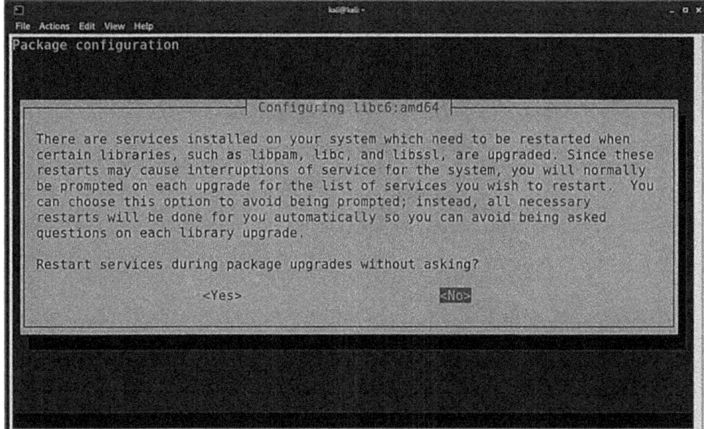

Figure 2.54 External libraries configuration.

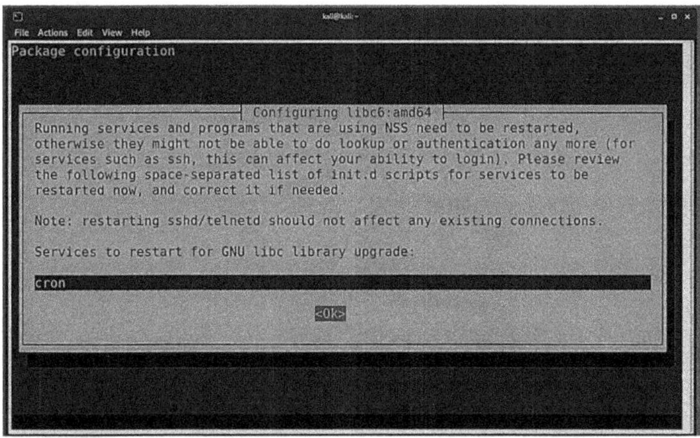

Figure 2.55 External libraries configuration messages.

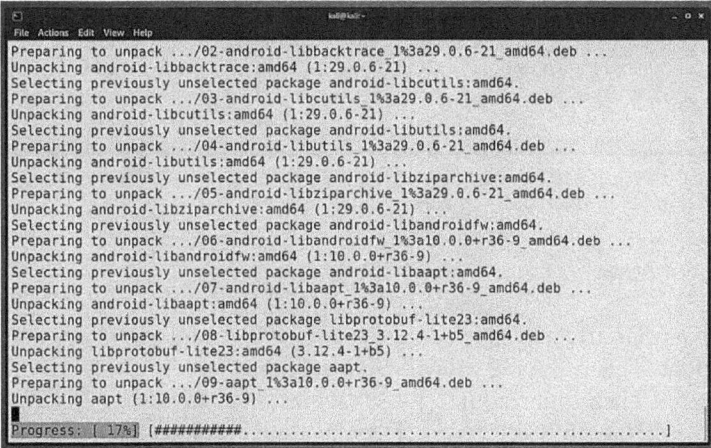

```
Preparing to unpack .../02-android-libbacktrace_1%3a29.0.6-21_amd64.deb ...
Unpacking android-libbacktrace:amd64 (1:29.0.6-21) ...
Selecting previously unselected package android-libcutils:amd64.
Preparing to unpack .../03-android-libcutils_1%3a29.0.6-21_amd64.deb ...
Unpacking android-libcutils:amd64 (1:29.0.6-21) ...
Selecting previously unselected package android-libutils:amd64.
Preparing to unpack .../04-android-libutils_1%3a29.0.6-21_amd64.deb ...
Unpacking android-libutils:amd64 (1:29.0.6-21) ...
Selecting previously unselected package android-libziparchive:amd64.
Preparing to unpack .../05-android-libziparchive_1%3a29.0.6-21_amd64.deb ...
Unpacking android-libziparchive:amd64 (1:29.0.6-21) ...
Selecting previously unselected package android-libandroidfw:amd64.
Preparing to unpack .../06-android-libandroidfw_1%3a10.0.0+r36-9_amd64.deb ...
Unpacking android-libandroidfw:amd64 (1:10.0.0+r36-9) ...
Selecting previously unselected package android-libaapt:amd64.
Preparing to unpack .../07-android-libaapt_1%3a10.0.0+r36-9_amd64.deb ...
Unpacking android-libaapt:amd64 (1:10.0.0+r36-9) ...
Selecting previously unselected package libprotobuf-lite23:amd64.
Preparing to unpack .../08-libprotobuf-lite23_3.12.4-1+b5_amd64.deb ...
Unpacking libprotobuf-lite23:amd64 (3.12.4-1+b5) ...
Selecting previously unselected package aapt.
Preparing to unpack .../09-aapt_1%3a10.0.0+r36-9_amd64.deb ...
Unpacking aapt (1:10.0.0+r36-9) ...

Progress: [ 17%] [###########.................................................]
```

Figure 2.56 Installation of apktool.

```
File Actions Edit View Help
┌──(kali㉿kali)-[~]
└─$ apktool --help
Picked up _JAVA_OPTIONS: -Dawt.useSystemAAFontSettings=on -Dswing.aatext=true
Unrecognized option: --help
Apktool v2.6.1-dirty - a tool for reengineering Android apk files
with smali v2.4.0-debian and baksmali v2.4.0-debian
Copyright 2010 Ryszard Wiśniewski <brut.alll@gmail.com>
Copyright 2010 Connor Tumbleson <connor.tumbleson@gmail.com>

usage: apktool
 -advance,--advanced   prints advance information.
 -version,--version    prints the version then exits
usage: apktool if|install-framework [options] <framework.apk>
 -p,--frame-path <dir>   Stores framework files into <dir>.
 -t,--tag <tag>          Tag frameworks using <tag>.
usage: apktool d[ecode] [options] <file_apk>
 -f,--force              Force delete destination directory.
 -o,--output <dir>       The name of folder that gets written. Default is apk.out
 -p,--frame-path <dir>   Uses framework files located in <dir>.
 -r,--no-res             Do not decode resources.
 -s,--no-src             Do not decode sources.
 -t,--frame-tag <tag>    Uses framework files tagged by <tag>.
usage: apktool b[uild] [options] <app_path>
 -f,--force-all          Skip changes detection and build all files.
```

Figure 2.57 apktool has been installed.

```
File Actions Edit View Help
openbabel                               openpaperwork-gtk
openbabel-gui                           openpaperwork-gtk-doc
openbgpd                                openpgp-applet
openboard-fonts-nonfree                 openpref
┌──(kali㉿kali)-[~]
└─$ sudo apt install openjdk-11-jre-headless
Reading package lists... Done
Building dependency tree... Done
Reading state information... Done
The following additional packages will be installed:
  openjdk-11-jre
Suggested packages:
  fonts-ipafont-gothic fonts-ipafont-mincho fonts-wqy-microhei | fonts-wqy-zenhei
  fonts-indic
The following packages will be upgraded:
  openjdk-11-jre openjdk-11-jre-headless
2 upgraded, 0 newly installed, 0 to remove and 1665 not upgraded.
Need to get 37.6 MB of archives.
After this operation, 385 kB of additional disk space will be used.
Do you want to continue? [Y/n] y
Get:1 http://http.kali.org/kali kali-rolling/main amd64 openjdk-11-jre amd64 11.0.16+8
-1 [175 kB]
Get:2 http://http.kali.org/kali kali-rolling/main amd64 openjdk-11-jre-headless amd64
11.0.16+8-1 [37.4 MB]
12% [2 openjdk-11-jre-headless 959 kB/37.4 MB 3%]
```

Figure 2.58 Java installation in Kali

For Windows:

Step #1: Go to PowerShell and type *winget search Java.*
Step #2: This is the file we will be installing.
Step #3: Then to install it, type *winget install Oracle.jdk.17.*
Step #4: Go to C:\Program Files\Java\jdk-17.0.4.1\bin and copy the location.

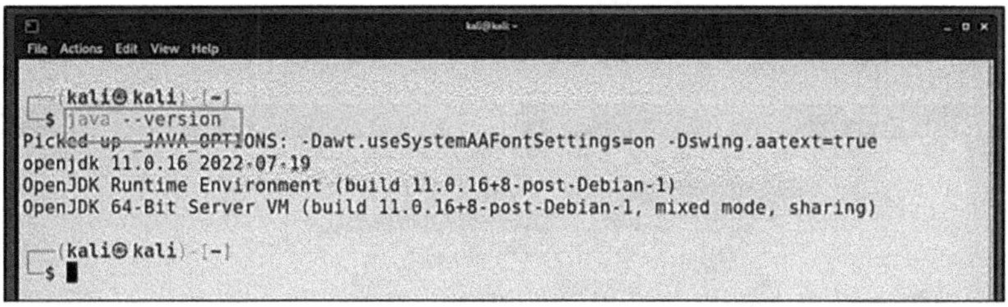

```
File  Actions  Edit  View  Help

┌──(kali㉿kali)-[~]
└─$ java --version
Picked up _JAVA_OPTIONS: -Dawt.useSystemAAFontSettings=on -Dswing.aatext=true
openjdk 11.0.16 2022-07-19
OpenJDK Runtime Environment (build 11.0.16+8-post-Debian-1)
OpenJDK 64-Bit Server VM (build 11.0.16+8-post-Debian-1, mixed mode, sharing)

┌──(kali㉿kali)-[~]
└─$
```

Figure 2.59 Checking java version.

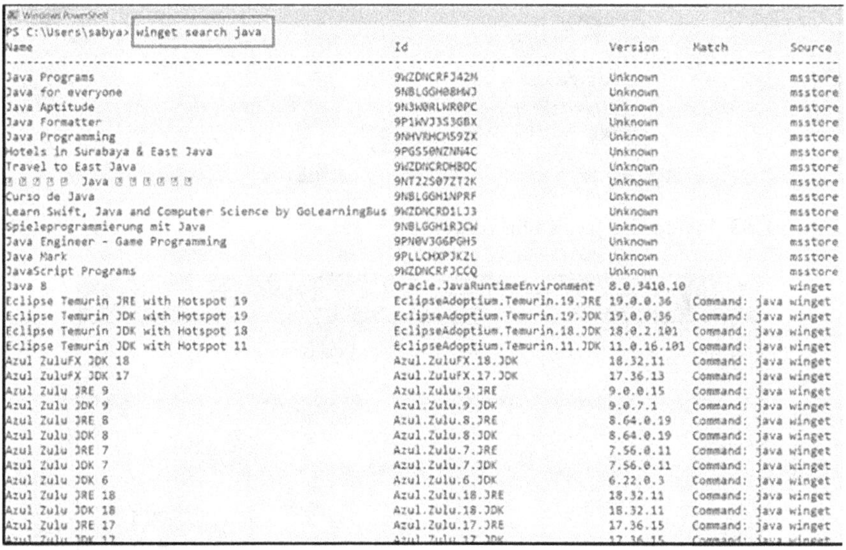

```
Windows PowerShell
PS C:\Users\sabya> winget search java
Name                                                Id                                 Version       Match          Source
---------------------------------------------------------------------------------------------------------------------------
Java Programs                                       9WZDNCRFJ42M                       Unknown                      msstore
Java for everyone                                   9NBLGGH08MWJ                       Unknown                      msstore
Java Aptitude                                       9N3W0RLWR0PC                       Unknown                      msstore
Java Formatter                                      9P1WVJ3S3GBX                       Unknown                      msstore
Java Programming                                    9NHVRHCM59ZX                       Unknown                      msstore
Hotels in Surabaya & East Java                      9PGS50NZNN4C                       Unknown                      msstore
Travel to East Java                                 9WZDNCRDHBDC                       Unknown                      msstore
□□□□□ Java □□□□□□                                   9NT22S07ZT2K                       Unknown                      msstore
Curso de Java                                       9NBLGGH1NPRF                       Unknown                      msstore
Learn Swift, Java and Computer Science by GoLearningBus 9WZDNCRD1LJ3                   Unknown                      msstore
Spieleprogrammierung mit Java                       9NBLGGH1RJCW                       Unknown                      msstore
Java Engineer - Game Programming                    9PN0V3G6PGH5                       Unknown                      msstore
Java Mark                                           9PLLCHXPJKZL                       Unknown                      msstore
JavaScript Programs                                 9WZDNCRFJCCQ                       Unknown                      msstore
Java 8                                              Oracle.JavaRuntimeEnvironment      8.0.3410.10                  winget
Eclipse Temurin JRE with Hotspot 19                 EclipseAdoptium.Temurin.19.JRE     19.0.0.36     Command: java  winget
Eclipse Temurin JDK with Hotspot 19                 EclipseAdoptium.Temurin.19.JDK     19.0.0.36     Command: java  winget
Eclipse Temurin JDK with Hotspot 18                 EclipseAdoptium.Temurin.18.JDK     18.0.2.101    Command: java  winget
Eclipse Temurin JDK with Hotspot 11                 EclipseAdoptium.Temurin.11.JDK     11.0.16.101   Command: java  winget
Azul ZuluFX JDK 18                                  Azul.ZuluFX.18.JDK                 18.32.11      Command: java  winget
Azul ZuluFX JDK 17                                  Azul.ZuluFX.17.JDK                 17.36.13      Command: java  winget
Azul Zulu JRE 9                                     Azul.Zulu.9.JRE                    9.0.0.15      Command: java  winget
Azul Zulu JDK 9                                     Azul.Zulu.9.JDK                    9.0.7.1       Command: java  winget
Azul Zulu JRE 8                                     Azul.Zulu.8.JRE                    8.64.0.19     Command: java  winget
Azul Zulu JDK 8                                     Azul.Zulu.8.JDK                    8.64.0.19     Command: java  winget
Azul Zulu JRE 7                                     Azul.Zulu.7.JRE                    7.56.0.11     Command: java  winget
Azul Zulu JDK 7                                     Azul.Zulu.7.JDK                    7.56.0.11     Command: java  winget
Azul Zulu JDK 6                                     Azul.Zulu.6.JDK                    6.22.0.3      Command: java  winget
Azul Zulu JRE 18                                    Azul.Zulu.18.JRE                   18.32.11      Command: java  winget
Azul Zulu JDK 18                                    Azul.Zulu.18.JDK                   18.32.11      Command: java  winget
Azul Zulu JRE 17                                    Azul.Zulu.17.JRE                   17.36.15      Command: java  winget
Azul Zulu JDK 17                                    Azul.Zulu.17.JDK                   17.36.15      Command: java  winget
```

Figure 2.60 Searching for java using winget.

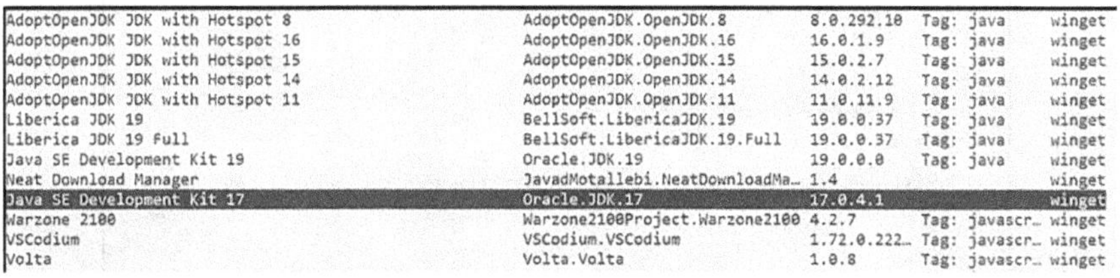

```
AdoptOpenJDK JDK with Hotspot 8                     AdoptOpenJDK.OpenJDK.8             8.0.292.10    Tag: java      winget
AdoptOpenJDK JDK with Hotspot 16                    AdoptOpenJDK.OpenJDK.16            16.0.1.9      Tag: java      winget
AdoptOpenJDK JDK with Hotspot 15                    AdoptOpenJDK.OpenJDK.15            15.0.2.7      Tag: java      winget
AdoptOpenJDK JDK with Hotspot 14                    AdoptOpenJDK.OpenJDK.14            14.0.2.12     Tag: java      winget
AdoptOpenJDK JDK with Hotspot 11                    AdoptOpenJDK.OpenJDK.11            11.0.11.9     Tag: java      winget
Liberica JDK 19                                     BellSoft.LibericaJDK.19            19.0.0.37     Tag: java      winget
Liberica JDK 19 Full                                BellSoft.LibericaJDK.19.Full       19.0.0.37     Tag: java      winget
Java SE Development Kit 19                          Oracle.JDK.19                      19.0.0.0      Tag: java      winget
Neat Download Manager                               JavadMotallebi.NeatDownloadMa…     1.4                          winget
Java SE Development Kit 17                          Oracle.JDK.17                      17.0.4.1                     winget
Warzone 2100                                        Warzone2100Project.Warzone2100     4.2.7         Tag: javascr…  winget
VSCodium                                            VSCodium.VSCodium                  1.72.0.222…   Tag: javascr…  winget
Volta                                               Volta.Volta                        1.0.8         Tag: javascr…  winget
```

Figure 2.61 The jdk we need to install.

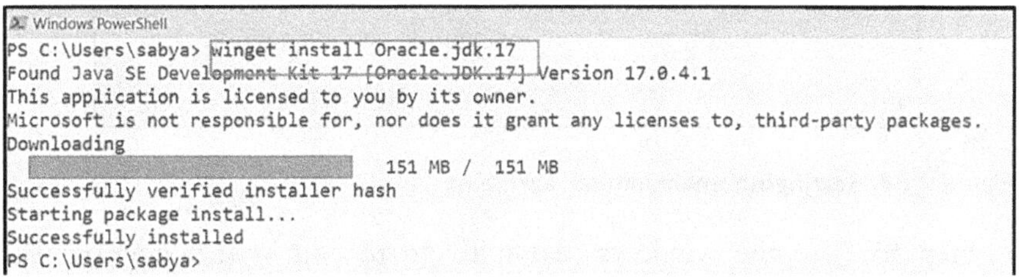

```
Windows PowerShell
PS C:\Users\sabya> winget install Oracle.jdk.17
Found Java SE Development Kit 17 [Oracle.JDK.17] Version 17.0.4.1
This application is licensed to you by its owner.
Microsoft is not responsible for, nor does it grant any licenses to, third-party packages.
Downloading
  ██████████████████████████████  151 MB /  151 MB
Successfully verified installer hash
Starting package install...
Successfully installed
PS C:\Users\sabya>
```

Figure 2.62 Jdk has been installed successfully.

bin

File Home Share View

C:\Program Files\Java\jdk-17.0.4.1\bin

Quick access: Desktop, Downloads, Documents, Pictures, kali_vbox, kali_vmware, Screenshots, tools; This PC: 3D Objects, Desktop

Name	Date modified	Type	Size
server	10-Oct-22 08:59 PM	File folder	
api-ms-win-core-console-l1-1-0.dll	10-Oct-22 08:59 PM	Application extens...	12 KB
api-ms-win-core-console-l1-2-0.dll	10-Oct-22 08:59 PM	Application extens...	12 KB
api-ms-win-core-datetime-l1-1-0.dll	10-Oct-22 08:59 PM	Application extens...	12 KB
api-ms-win-core-debug-l1-1-0.dll	10-Oct-22 08:59 PM	Application extens...	12 KB
api-ms-win-core-errorhandling-l1-1-0.dll	10-Oct-22 08:59 PM	Application extens...	12 KB
api-ms-win-core-file-l1-1-0.dll	10-Oct-22 08:59 PM	Application extens...	15 KB
api-ms-win-core-file-l1-2-0.dll	10-Oct-22 08:59 PM	Application extens...	12 KB
api-ms-win-core-file-l2-1-0.dll	10-Oct-22 08:59 PM	Application extens...	12 KB
api-ms-win-core-handle-l1-1-0.dll	10-Oct-22 08:59 PM	Application extens...	12 KB
api-ms-win-core-heap-l1-1-0.dll	10-Oct-22 08:59 PM	Application extens...	12 KB
api-ms-win-core-interlocked-l1-1-0.dll	10-Oct-22 08:59 PM	Application extens...	12 KB
api-ms-win-core-libraryloader-l1-1-0.dll	10-Oct-22 08:59 PM	Application extens...	13 KB

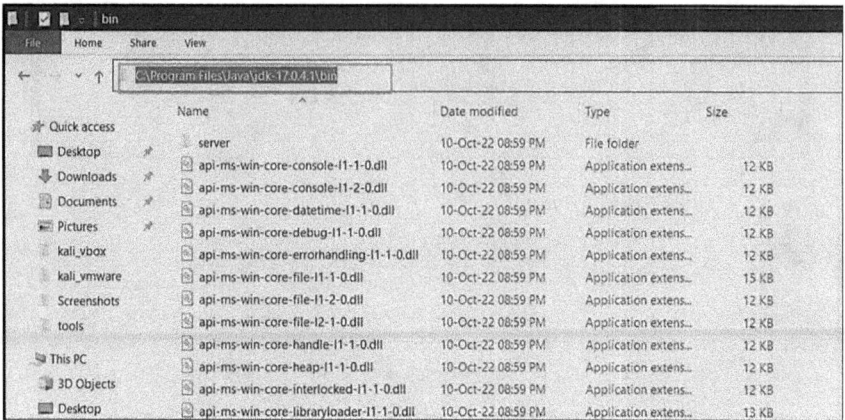

Figure 2.63 Inside the Java's bin/ folder.

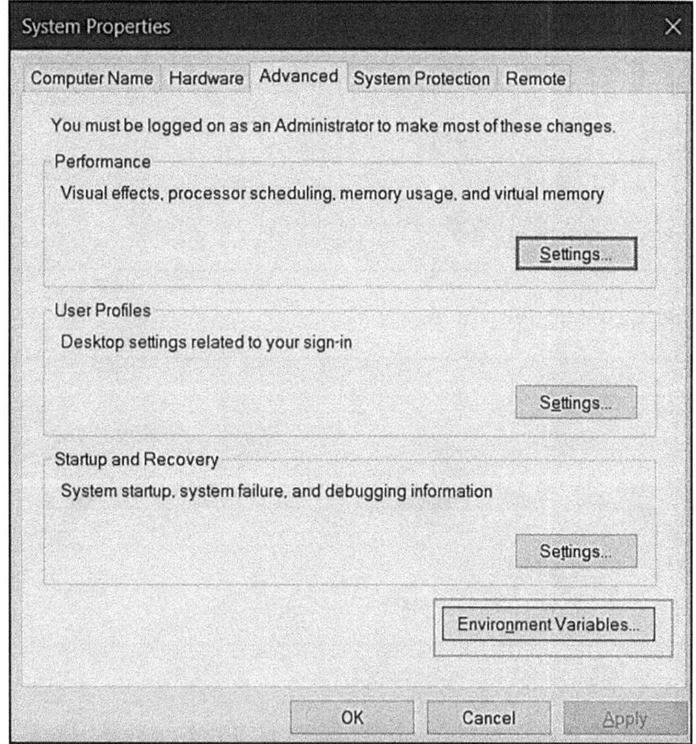

Figure 2.64 Setting up environment variables.

Step #5: Use the windows search to go to edit system environment variables.

Step #6: Then, click on environment variables > system variables > Path > Edit.

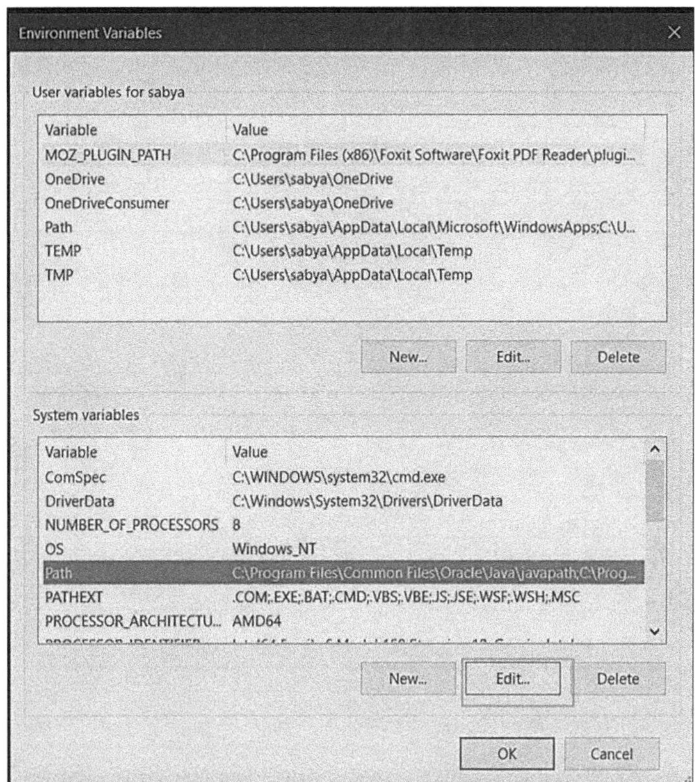

Figure 2.65 Editing system variables.

Step #7: Then, click New > Add the path > click OK.
Step #8: Now go to CMD or PowerShell and type *java – version* to confirm.

2.4.4 Installing keytool

Now since we already install Java, keytool gets automatically shipped with the Java Development Kit (JDK).

So, typing *keytool – help* would confirm the proper installation of keytool.

2.4.5 Installing adb

Step #1: Type *sudo apt install google-android-platform-tools-installer* in the Kali terminal.
Step #2: Type *adb – version* to confirm the installation.

2.4.6 Installing dex2jar

Step #1: Open the Kali terminal and type *sudo apt install dex2jar*.
Step #2: Type *d2j-dex2jar – help* for confirmation of proper installation.

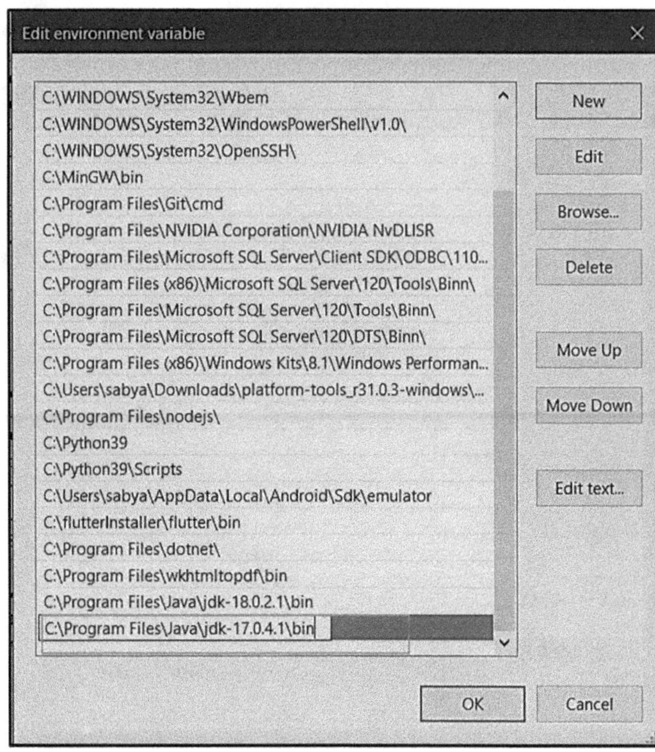

Figure 2.66 Adding the path.

```
PS C:\Users\sabya> java --version
java 17.0.4.1 2022-08-18 LTS
Java(TM) SE Runtime Environment (build 17.0.4.1+1-LTS-2)
Java HotSpot(TM) 64-Bit Server VM (build 17.0.4.1+1-LTS-2, mixed mode, sharing)
PS C:\Users\sabya> _
```

Figure 2.67 Java has been installed successfully.

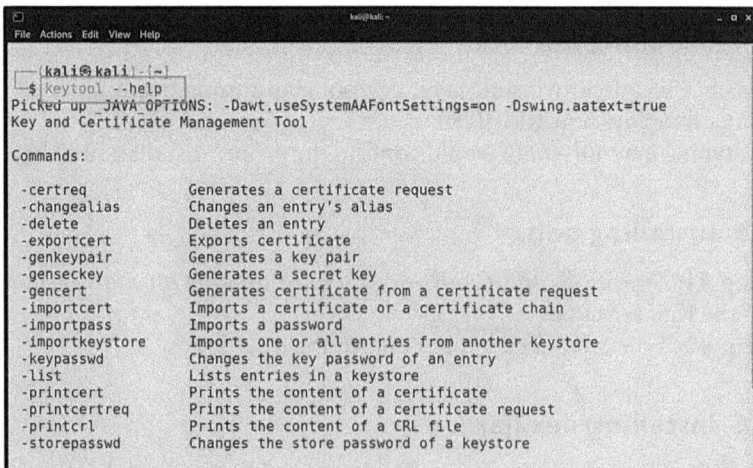

```
File Actions Edit View Help

┌──(kali㉿kali)-[~]
└─$ keytool --help
Picked up _JAVA_OPTIONS: -Dawt.useSystemAAFontSettings=on -Dswing.aatext=true
Key and Certificate Management Tool

Commands:

 -certreq             Generates a certificate request
 -changealias         Changes an entry's alias
 -delete              Deletes an entry
 -exportcert          Exports certificate
 -genkeypair          Generates a key pair
 -genseckey           Generates a secret key
 -gencert             Generates certificate from a certificate request
 -importcert          Imports a certificate or a certificate chain
 -importpass          Imports a password
 -importkeystore      Imports one or all entries from another keystore
 -keypasswd           Changes the key password of an entry
 -list                Lists entries in a keystore
 -printcert           Prints the content of a certificate
 -printcertreq        Prints the content of a certificate request
 -printcrl            Prints the content of a CRL file
 -storepasswd         Changes the store password of a keystore
```

Figure 2.68 keytool has been installed successfully.

```
kali@kali: ~
File  Actions  Edit  View  Help

┌──(kali㉿kali)-[~]
└─$ sudo apt install google-android-platform-tools-installer
Reading package lists... Done
Building dependency tree... Done
Reading state information... Done
The following NEW packages will be installed:
  google-android-platform-tools-installer
0 upgraded, 1 newly installed, 0 to remove and 1664 not upgraded.
Need to get 20.9 kB of archives.
After this operation, 18.9 MB of additional disk space will be used.
Get:1 http://http.kali.org/kali kali-rolling/non-free amd64 google-android-platform-to
ols-installer amd64 33.0.2+1659100756 [20.9 kB]
Fetched 20.9 kB in 2s (9,527 B/s)
Preconfiguring packages ...
Selecting previously unselected package google-android-platform-tools-installer.
(Reading database ... 269607 files and directories currently installed.)
Preparing to unpack .../google-android-platform-tools-installer_33.0.2+1659100756_amd6
4.deb ...
Unpacking google-android-platform-tools-installer (33.0.2+1659100756) ...
Setting up google-android-platform-tools-installer (33.0.2+1659100756) ...
make: Entering directory '/var/cache/google-android-platform-tools-installer'
cd /var/cache/google-android-platform-tools-installer && \
        su nobody -s /bin/sh -c "wget --continue https://dl.google.com/android/reposit
ory/platform-tools_r33.0.2-linux.zip -O platform-tools_r33.0.2-linux.zip.tmp && mv pla
```

Figure 2.69 Android platform tools are getting installed.

```
kali@kali: ~
File  Actions  Edit  View  Help

┌──(kali㉿kali)-[~]
└─$ adb --version
Android Debug Bridge version 1.0.41
Version 33.0.2-8557947
Installed as /usr/lib/android-sdk/platform-tools/adb

┌──(kali㉿kali)-[~]
└─$ ▮
```

Figure 2.70 adb got installed successfully.

```
kali@kali: ~
File  Actions  Edit  View  Help

┌──(kali㉿kali)-[~]
└─$ sudo apt install dex2jar
Reading package lists... Done
Building dependency tree... Done
Reading state information... Done
The following NEW packages will be installed:
  dex2jar
0 upgraded, 1 newly installed, 0 to remove and 1664 not upgraded.
Need to get 4,985 kB of archives.
After this operation, 6,083 kB of additional disk space will be used.
Get:1 http://http.kali.org/kali kali-rolling/main amd64 dex2jar all 2.1~nightly-28-0ka
li2 [4,985 kB]
9% [1 dex2jar 558 kB/4,985 kB 11%]▮
```

Figure 2.71 Installing dex2jar in Kali.

```
File Actions Edit View Help
└─$ d2j-dex2jar --help
Picked up _JAVA_OPTIONS: -Dawt.useSystemAAFontSettings=on -Dswing.aatext=true
d2j-dex2jar -- convert dex to jar
usage: d2j-dex2jar [options] <file0> [file1 ... fileN]
options:
 --skip-exceptions              skip-exceptions
 -d,--debug-info                translate debug info
 -e,--exception-file <file>     detail exception file, default is $current_dir/[fi
                                le-name]-error.zip
 -f,--force                     force overwrite
 -h,--help                      Print this help message
 -n,--not-handle-exception      not handle any exceptions thrown by dex2jar
 -nc,--no-code
 -o,--output <out-jar-file>     output .jar file, default is $current_dir/[file-na
                                me]-dex2jar.jar
 -os,--optmize-synchronized     optimize-synchronized
 -p,--print-ir                  print ir to System.out
 -r,--reuse-reg                 reuse register while generate java .class file
 -s                             same with --topological-sort/-ts
 -ts,--topological-sort         sort block by topological, that will generate more
                                 readable code, default enabled
version: reader-2.1-SNAPSHOT, translator-2.1-SNAPSHOT, ir-2.1-SNAPSHOT

┌──(kali㉿kali)-[~]
└─$ ▮
```

Figure 2.72 dex2jar has been installed successfully.

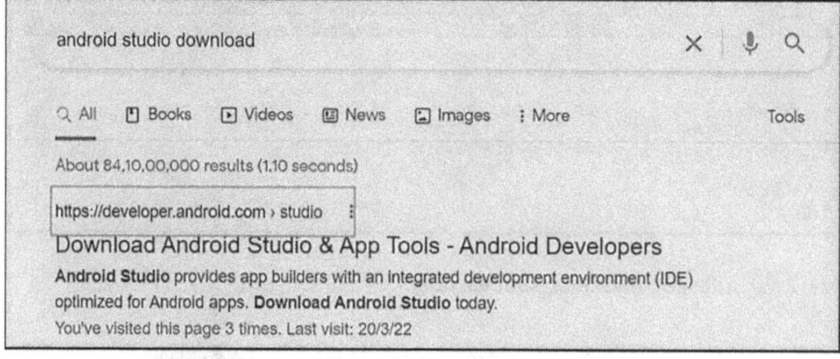

Figure 2.73 Searching for Android Studio in Google.

2.4.7 Installing Android Studio

Step 1: Go to Google, search for Android Studio, and click on the highlighted link.

Step 2: Click on the Download button, scroll down, and check the T&C agreement box.

Step 3: Your Android Studio will begin to download.

If you have any other system other than Windows, then you can select your preferred file

Step 4: Once downloaded, go to the explorer and open the executable. This will be shown.

Step 5: Wait for the load and start clicking Next, that's easy Windows is

Figure 2.74 Official download site of Android Studio.

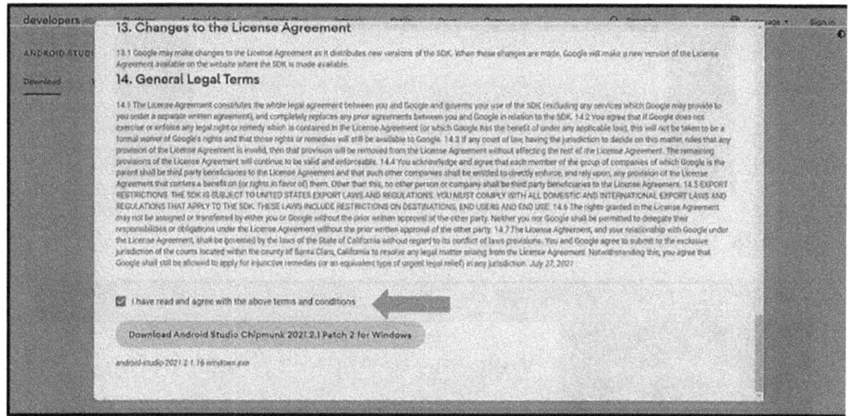

Figure 2.75 Terms & Condition agreement box.

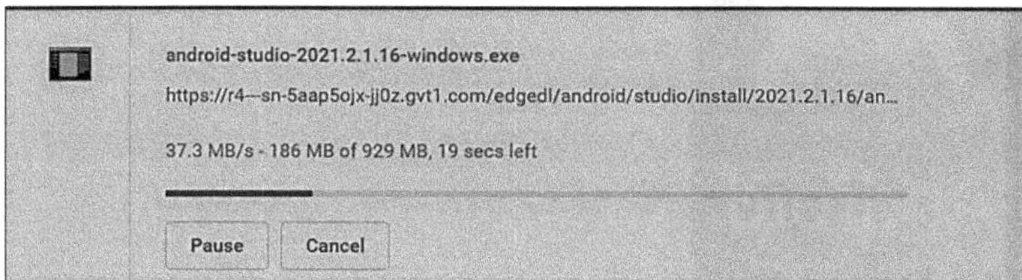

Figure 2.76 Downloading of Android Studio.

Step 6: Next, choose the Android Virtual device checkbox and click Next.
Step 7: Choose your installation path and click Next.
Step 8: Click Next again or check that box below if you don't want desktop shortcut.

Android Studio downloads

Platform	Android Studio package	Size	SHA-256 checksum
Windows (64-bit)	android-studio-2021.2.1.16-windows.exe Recommended	929 MiB	214fc733906090d615bb02f2576474a3d6c152249fb67b03124162e111da4c7
	android-studio-2021.2.1.16-windows.zip No .exe installer	940 MiB	78375a5b779d66630b757a32df0a26b8c052995623d649ed31787fb64eef14b1
Mac (64-bit)	android-studio-2021.2.1.16-mac.dmg	1017 MiB	df46f2199fcf4c7e6b882ba16151ea1d2dd48a15f5c87d30274f1b5401d2b648
Mac (64-bit, ARM)	android-studio-2021.2.1.16-mac_arm.dmg	1014 MiB	d4e06bcc6f614cd4b261fc6034529edb205b31b0e56824490a91350c3640806a
Linux (64-bit)	android-studio-2021.2.1.16-linux.tar.gz	964 MiB	aa5773a9e1da25bdb2367a8bdd2b623dbe0345170ed231e15b3140e8888447dc

Figure 2.77 Android Studio for another OS.

Figure 2.78 Installation prompt.

Figure 2.79 Android Studio installation setup.

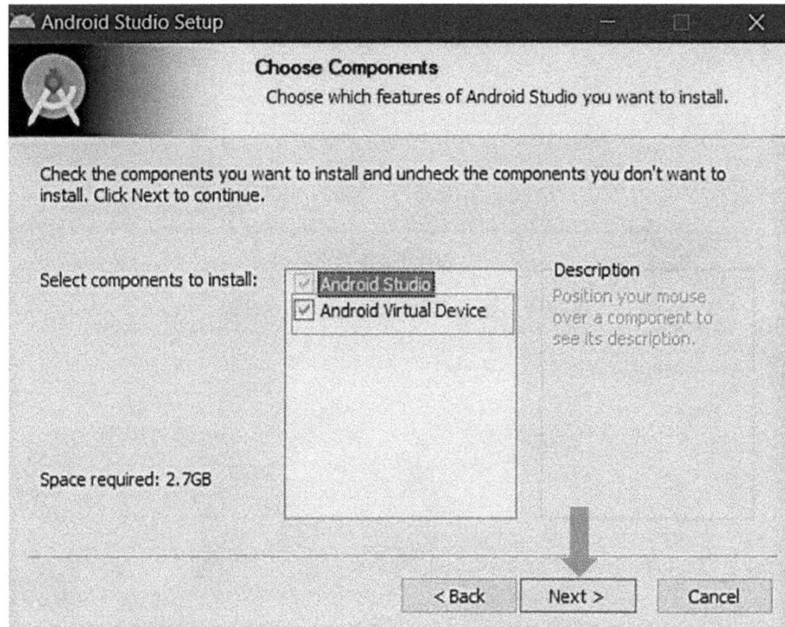

Figure 2.80 Additional component selection prompt.

Figure 2.81 Installation path of Android Studio.

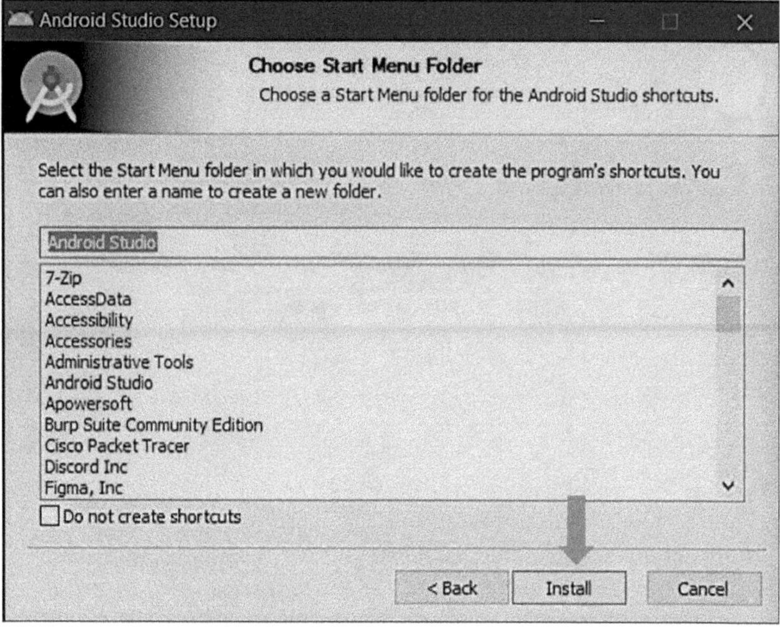

Figure 2.82 Start menu folder naming option.

Step 9: Your installation will begin, wait for the moment after finishing. Click Next.
Step 10: Click Finish and done. Enjoy!

Android Studio has been installed now. Let's start making the emulator.

2.4.8 Making the Android Emulator

Alright, so from here, the easy yet lengthy process begins. So, fasten your seatbelts, it's going to be an exciting journey.

Step #1: Go to the Start button and open Android Studio.
Step #2: On the bottom-right corner, click Emulator.
Step #3: Click on the Device Manager.
Step #4: Click on the Create Virtual device.
Step #5: If nothing happens despite clicking several times, click on the SDK Manager on the top-right corner (the center one with box and down arrow).
Step #6: Choose Android API 33, if not getting selected, click Edit.
Step #7: Click Next, make sure your API 33 is selected
Step #8: Click Next, accept the agreement, and click Next to start the installation of SDK.

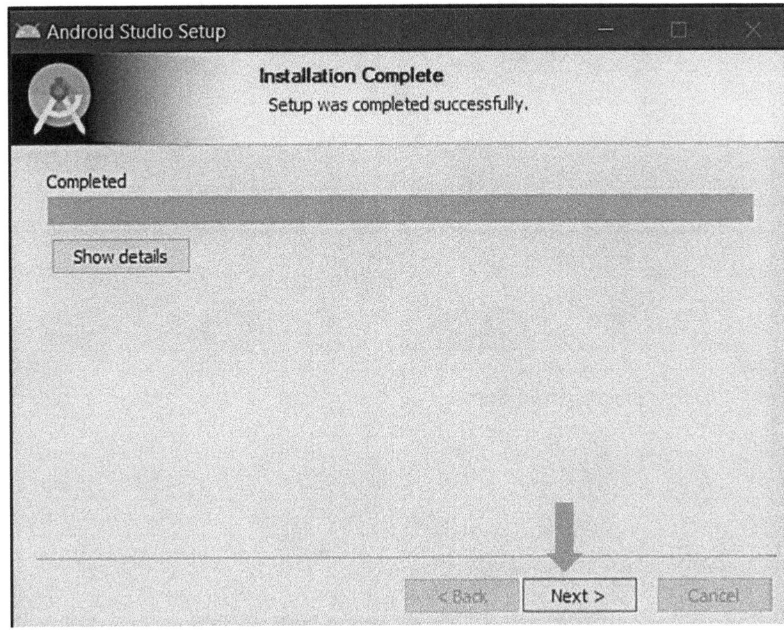

Figure 2.83 Installation of Android Studio.

Figure 2.84 Installation completed.

Figure 2.85 Starting Android Studio.

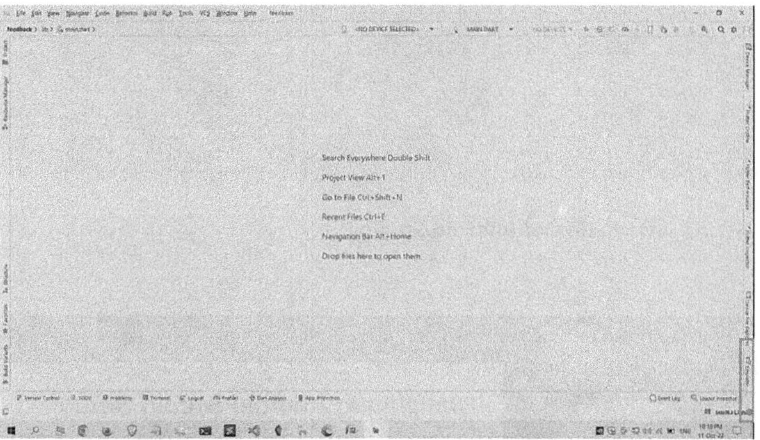

Figure 2.86 The Android Studio window.

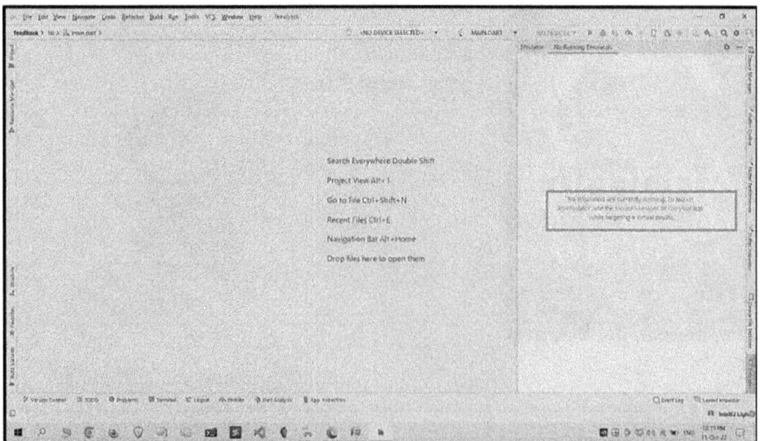

Figure 2.87 The emulator pane in Android Studio.

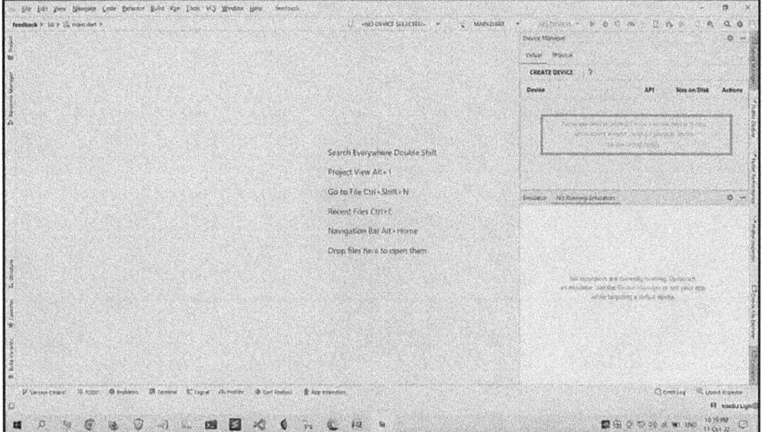

Figure 2.88 The Create Virtual device pane in Android Studio.

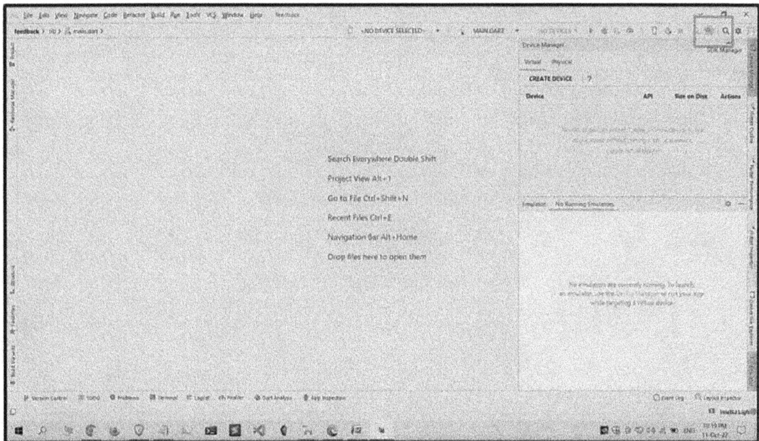

Figure 2.89 Android SDK Manager icon in Android Studio.

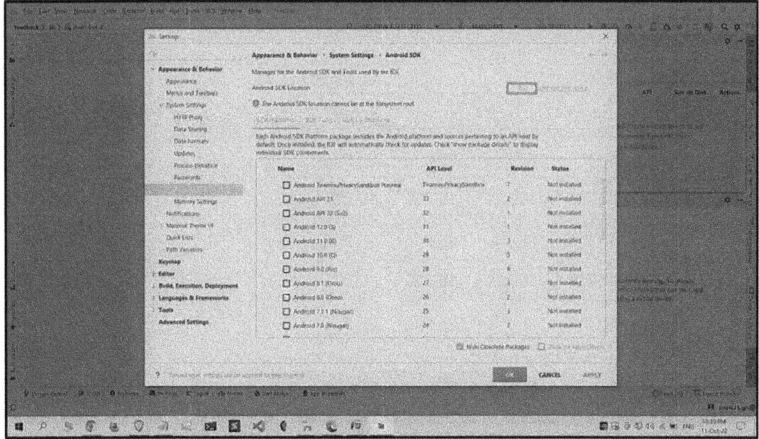

Figure 2.90 Settings window in Android Studio.

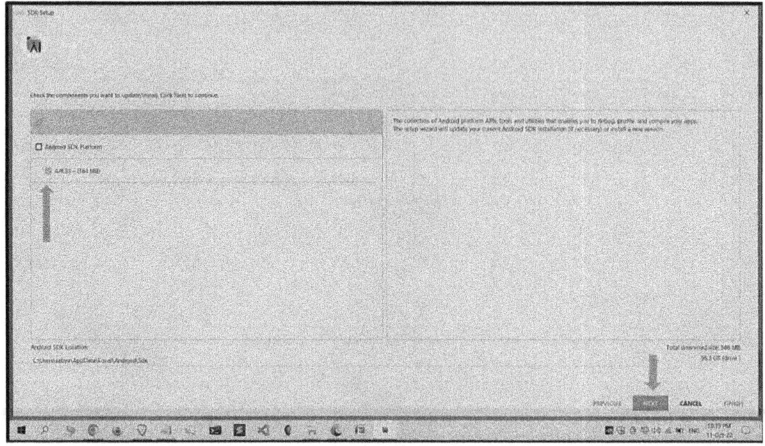

Figure 2.91 Android SDK Installation.

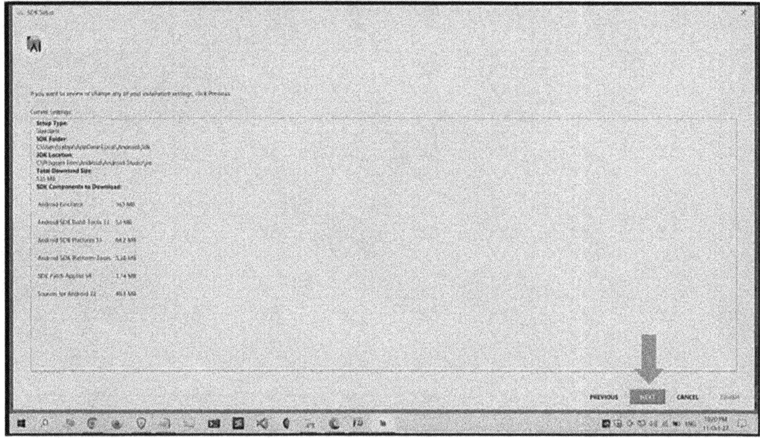

Figure 2.92 Android SDK Installation confirmation.

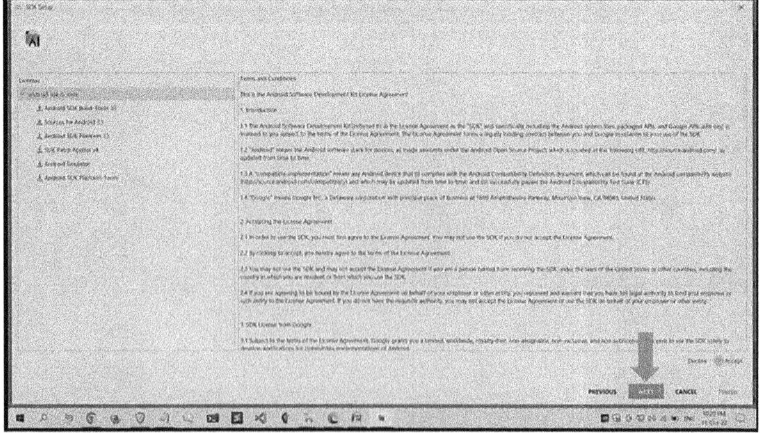

Figure 2.93 Android SDK Installation T&C.

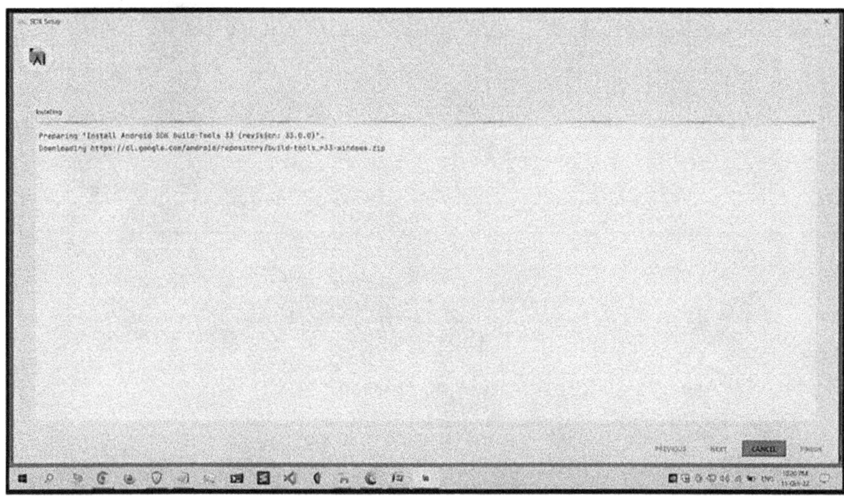

Figure 2.94 Android SDK Installation.

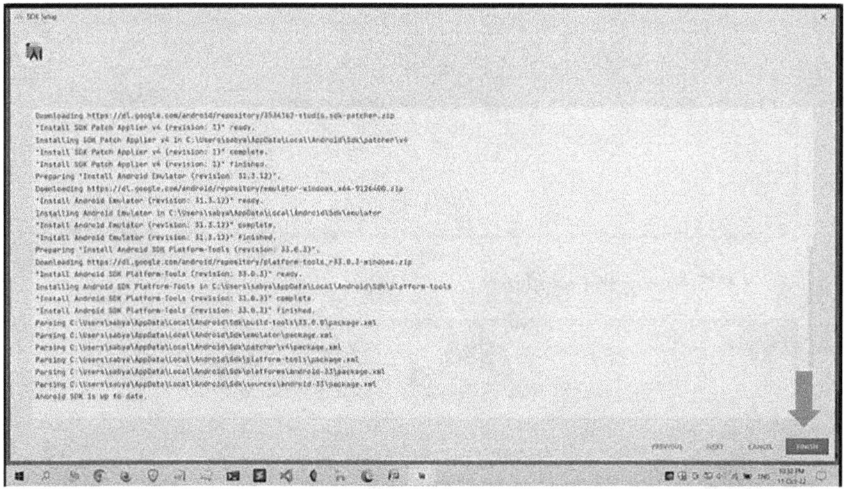

Figure 2.95 Android SDK Installation finished.

Once done, click Finish. Your Android SDK should be checked and then click OK.

Step #9: Wait for the above installation to finish and now click on the Create Virtual device.

Step #10: Choose Pixel XL 5.5″ as your emulator and click Next.

Step #11: Choose the system image as Oreo of API 27 x86

Step #12: Click Download to start downloading the system image shortly

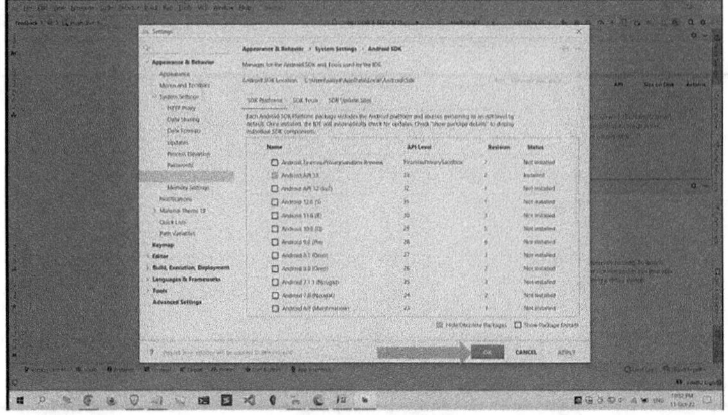

Figure 2.96 Android SDK Installation completed.

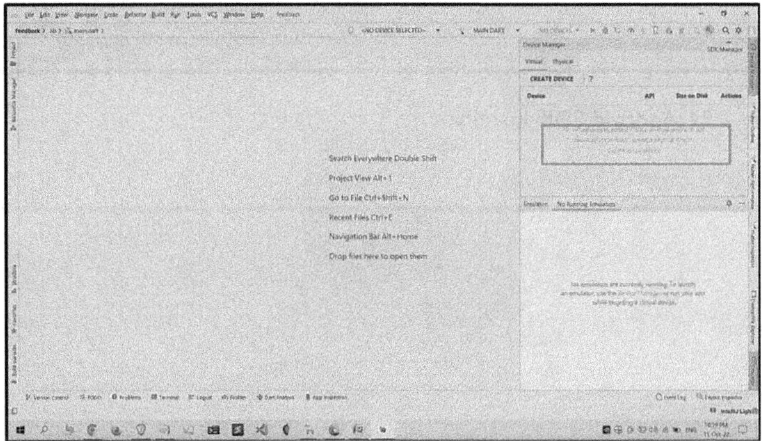

Figure 2.97 Creating the emulator.

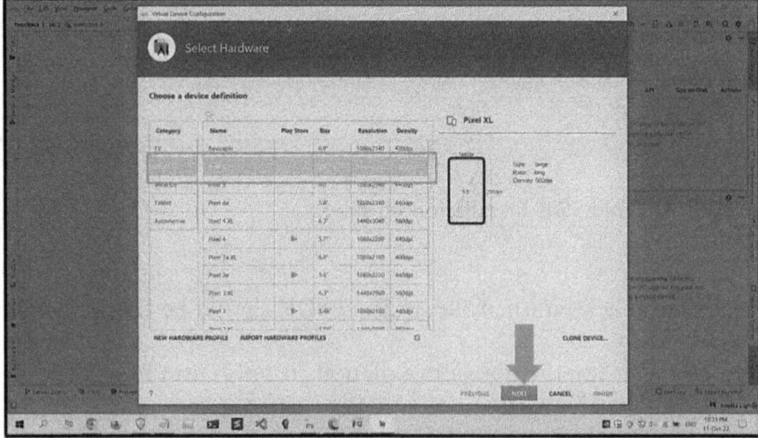

Figure 2.98 Emulator device selection.

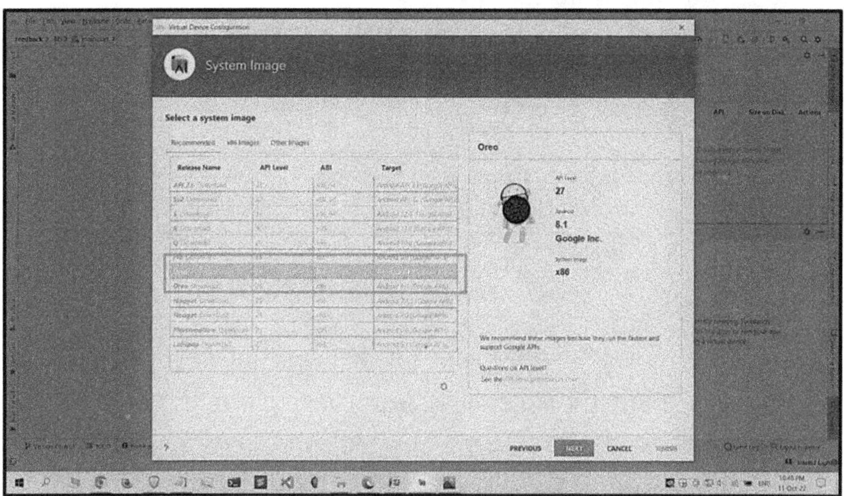

Figure 2.99 Android System Image selection.

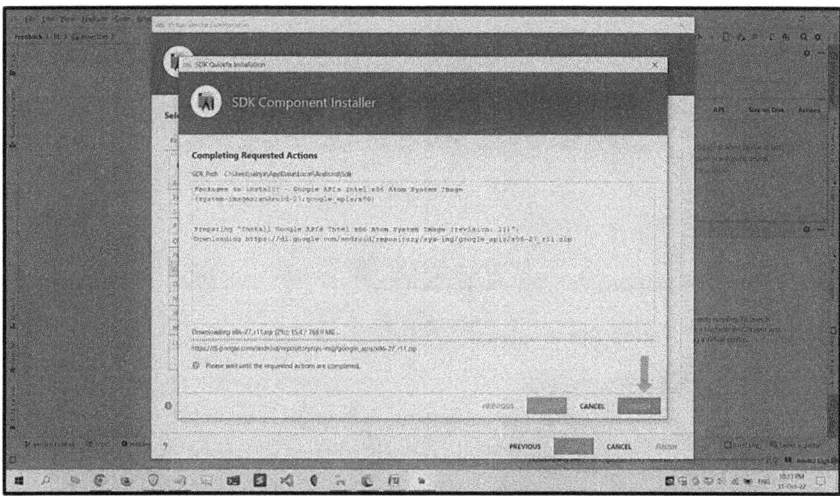

Figure 2.100 System Image installation.

Step #13: After it is downloaded, click Finish.

Step #14: Select the downloaded system image again and click Next.

Step #15: Here you can rename your device if needed and click Finish.

Step #16: In the window, you can see the emulator listed. Click on the Play button, the emulator will start in the window below, and you can resize to maximize the window size to increase the screen size of your phone.

Step #17: Click on the Power button to turn off the emulator. Feel free to explore other options as well.

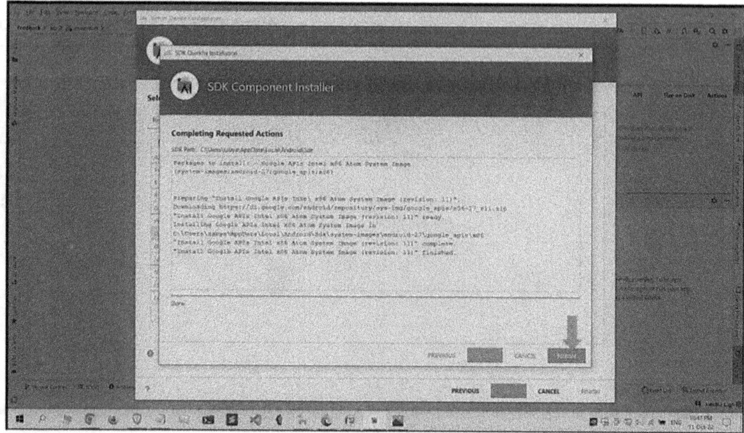

Figure 2.101 Android SDK Installation completed.

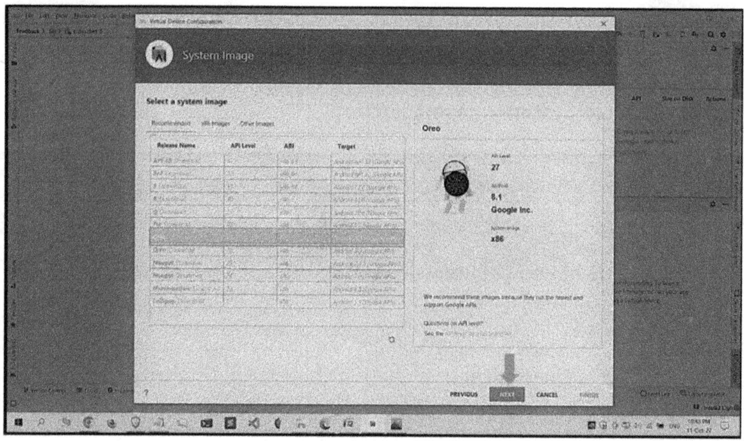

Figure 2.102 Creating the emulator.

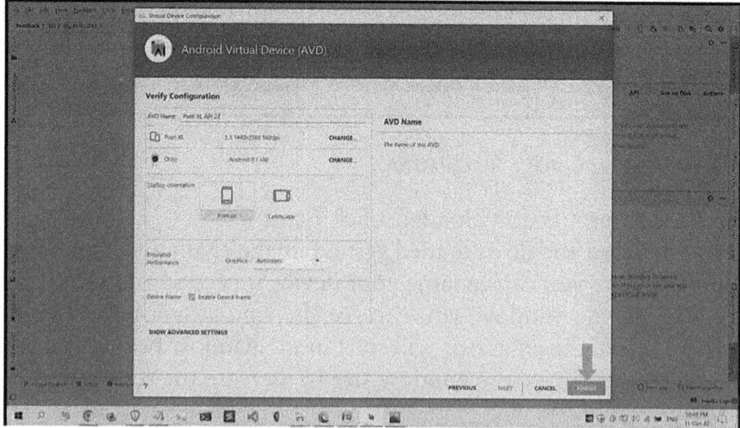

Figure 2.103 Final setup of Android emulator.

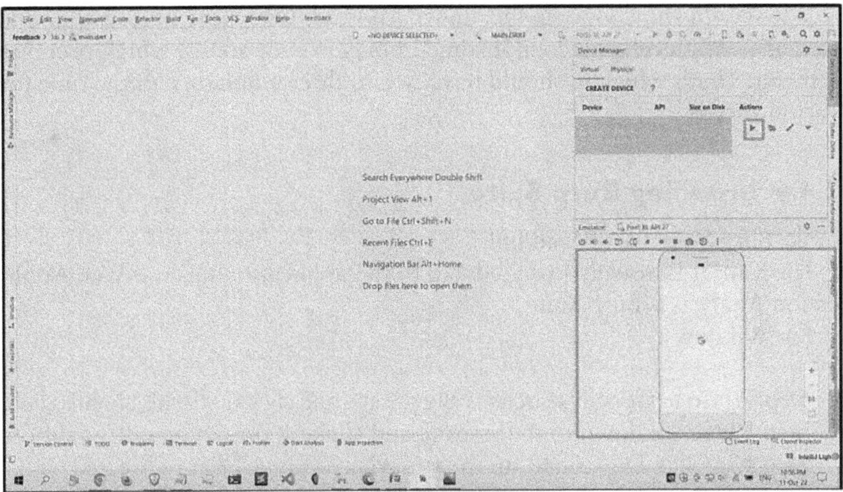

Figure 2.104 Android emulator booting up.

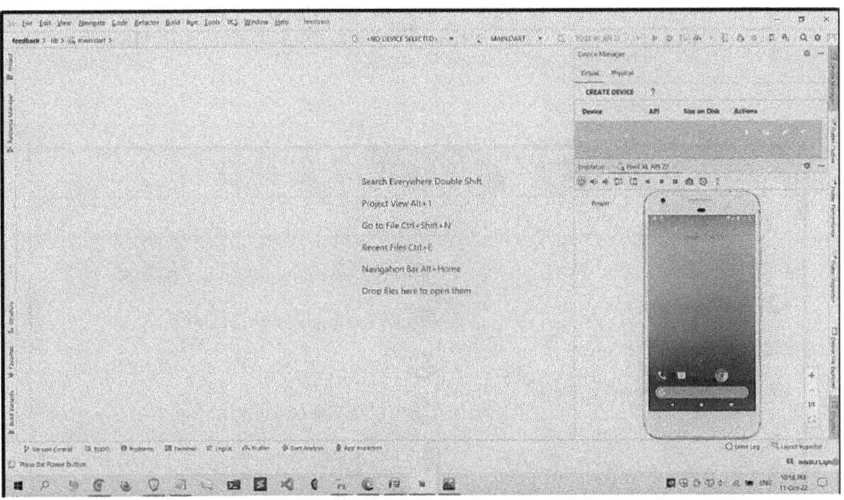

Figure 2.105 Android emulator running successfully.

Important: As a pentester you should have at least two to three emulators for different purposes. Let me explain. Due to different API levels, in some emulators you get Google Play Store installed and in some you don't. So, to test apps in Play Store, you need a system image of the emulator of API 29 or 30. I've personally used the image Q (present in Android Studio), which is API 29. Also make sure that when you select the image, on the right-hand side it should show Google Inc. under Android. One more need is that due to higher API level, you cannot sometimes gain root access, which becomes difficult. That's why you should have two to three emulators all the time for various needs.

2.4.9 Installing Burp Suite

Link: Burp Suite Download link

For Kali: It is already installed. Go to the application menu > Web Application Analysis > Burp Suite

For Windows:

Step #1: Go to Google, search for Burp Suite, and click on the highlighted link.
Step #2: Scroll down to the website and click on the Community Edition.
Step #3: Click Download on the site where you've been redirected.
Step #4: Your download will start.
Step #5: After it gets downloaded, open the explorer and execute it.
If you get this prompt, click Run.
Step #6: Start clicking Next and finally Finish at last.
Step #7: After installation, open Burp Suite using a start menu or a desktop shortcut. First select Temporary Project, click Next, and then Finish.

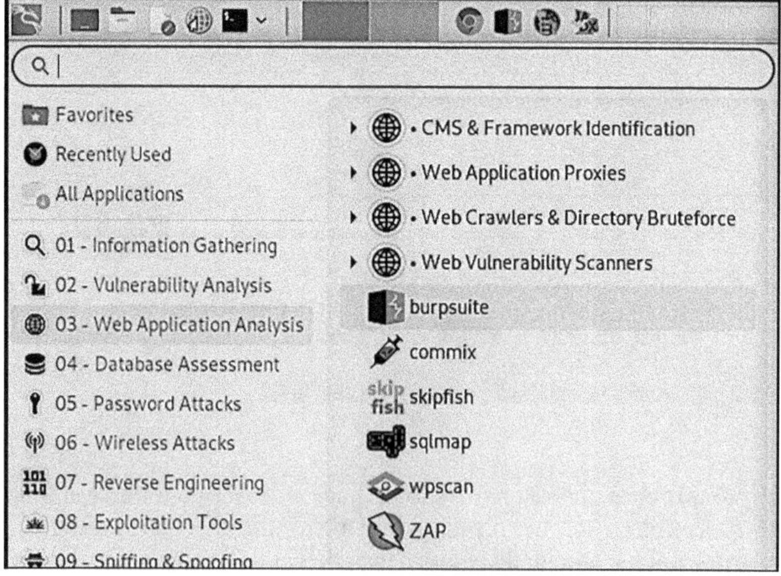

Figure 2.106 Location of Burp Suite in Kali.

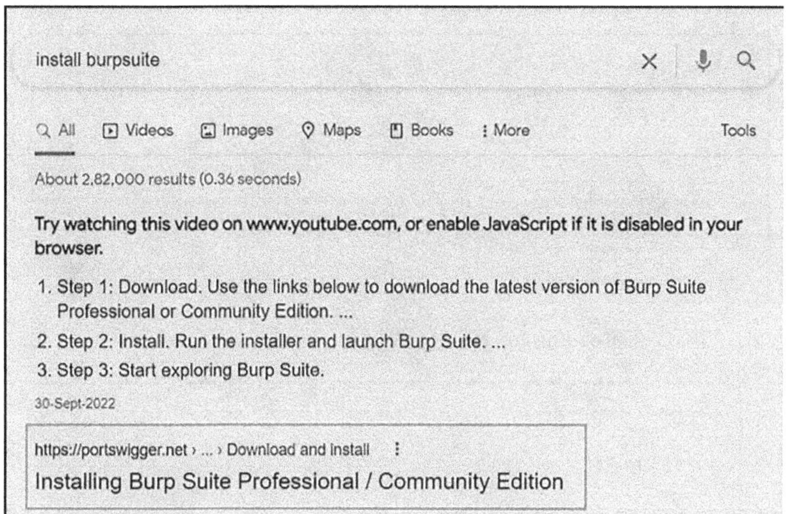

Figure 2.107 Searching for Burp Suite in Google.

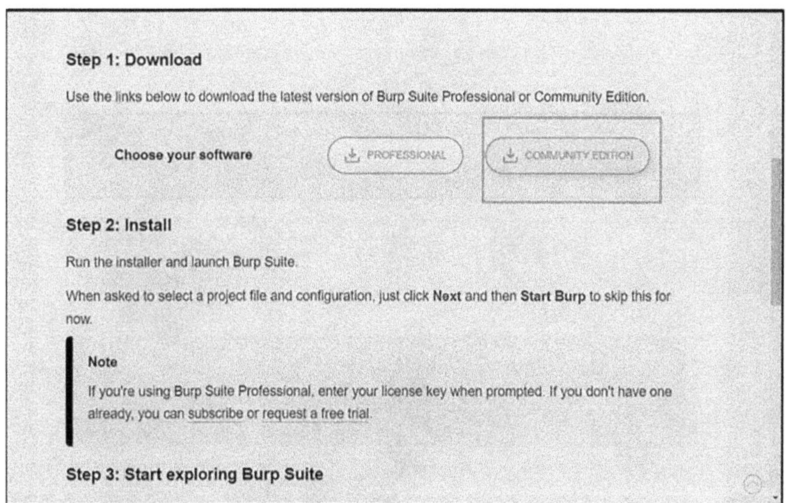

Figure 2.108 Official site of Burp Suite.

Look how beautiful she is. If you don't comprehend it, don't worry. For novices, this software can be a little challenging, but I'll make it incredibly simple for you. At first, utilizing this software was challenging for me as well, but it is now my main resource.

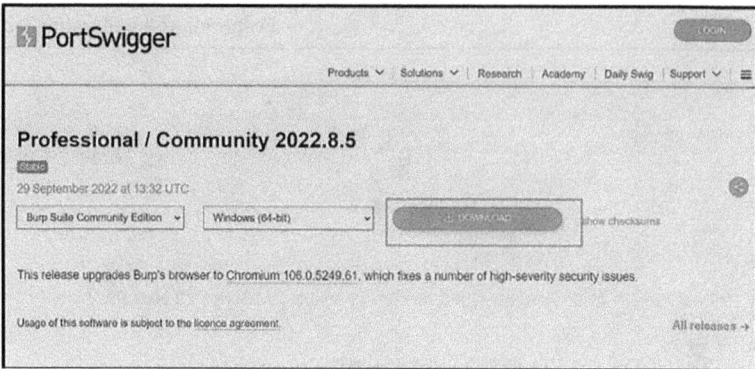

Figure 2.109 Downloading Burp Suite.

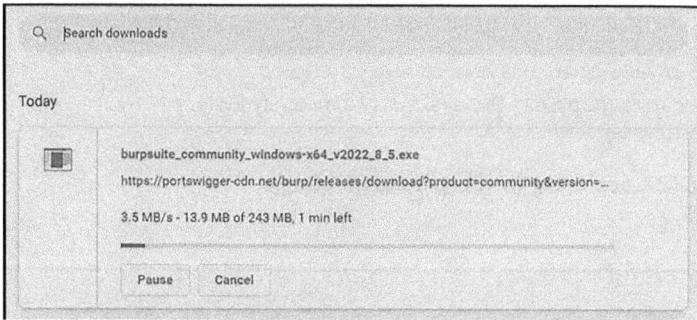

Figure 2.110 Download prompt.

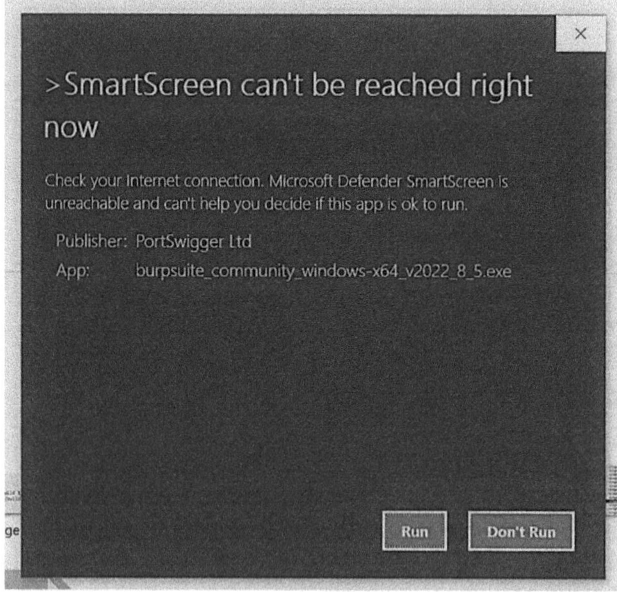

Figure 2.111 SmartScreen window.

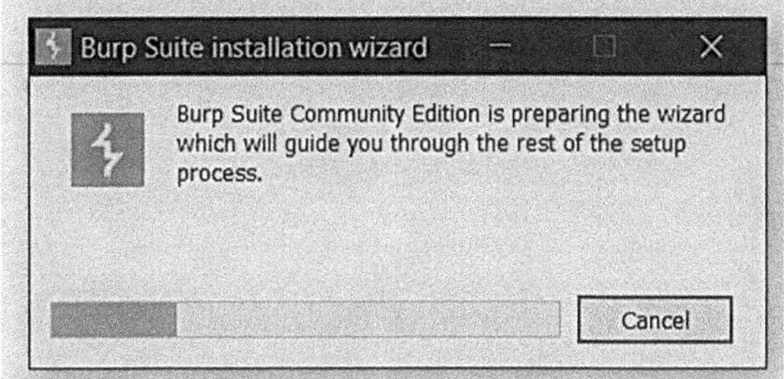

Figure 2.112 Installation Wizard.

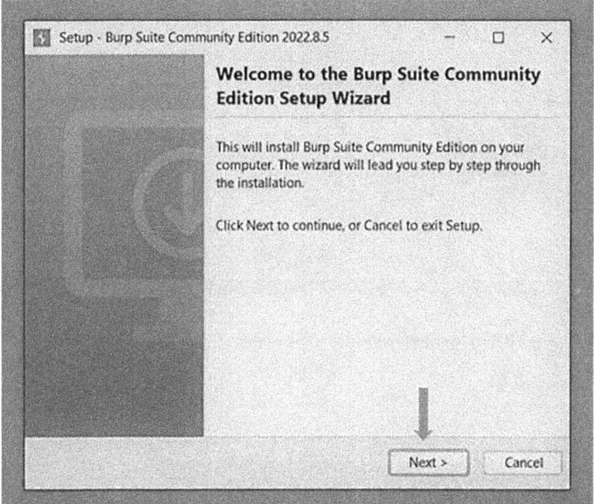

Figure 2.113 Burp Suite setup wizard.

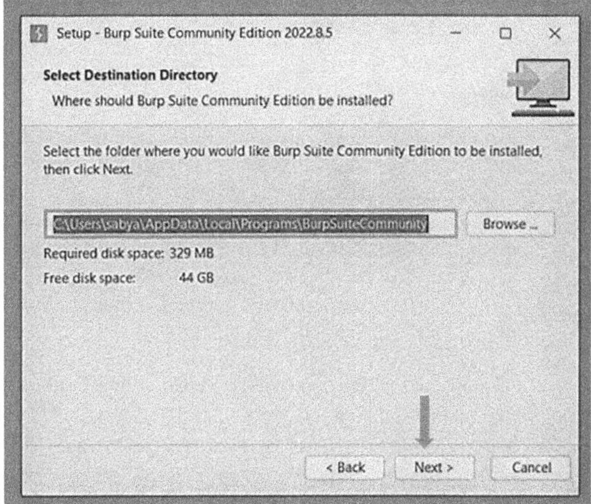

Figure 2.114 Burp Suite installation path.

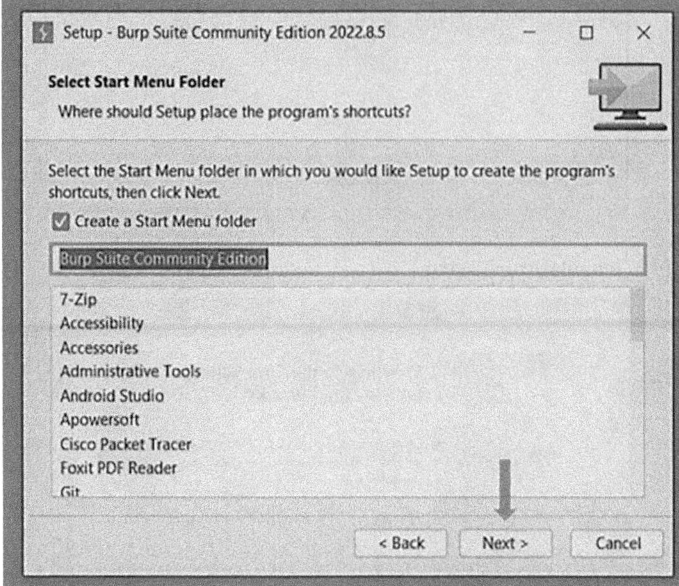

Figure 2.115 Burp Suite start menu folder configuration.

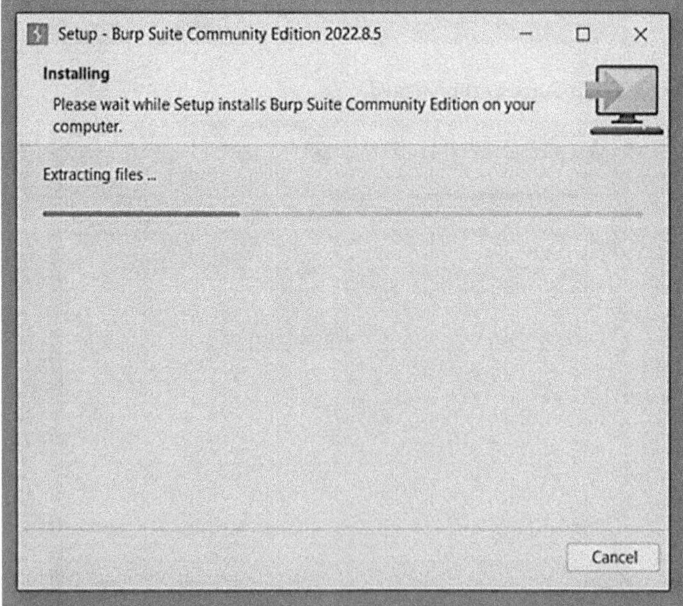

Figure 2.116 Burp Suite setup being installed.

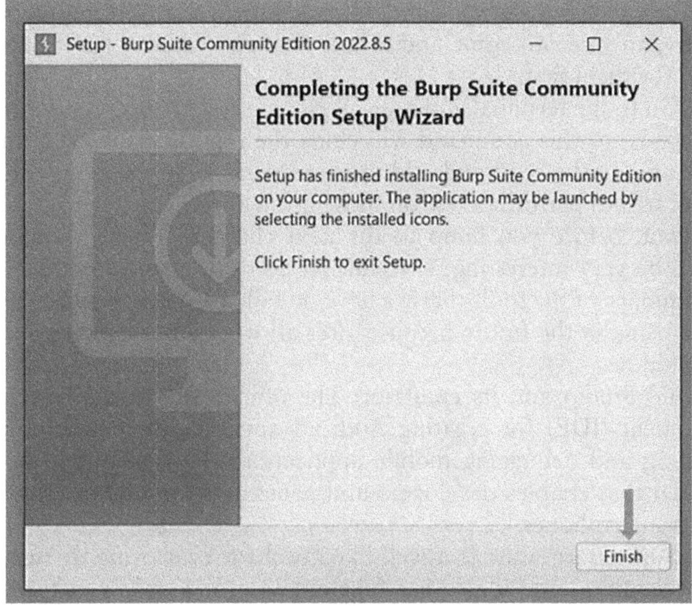

Figure 2.117 Burp Suite installed successfully.

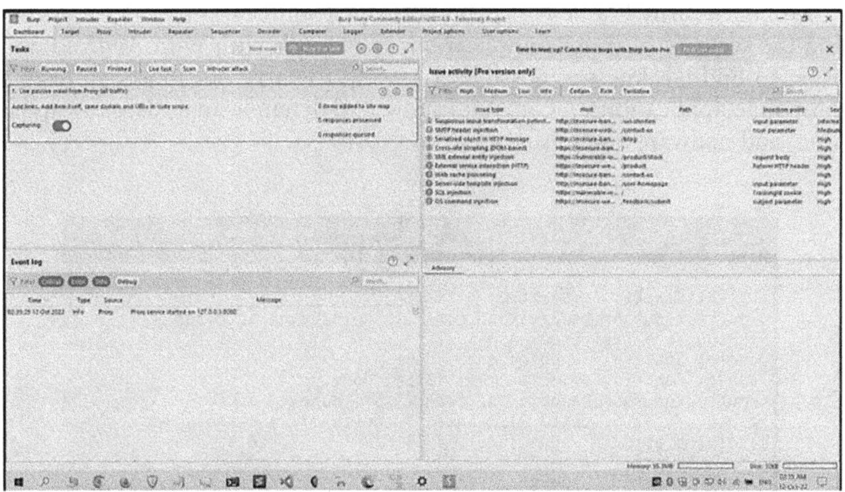

Figure 2.118 Burp Suite window.

2.4.10 Downloading the Vulnerable Apps

The testing will be done on applications that have been purposefully left open to attack. All the apps can be downloaded from their respective websites, which are all accessible via the Internet, but for ease of access, I've already downloaded the apks and uploaded them to my Github repository. This way, anyone can come and install the apps without having to search the web for them first.

Step: Go to the terminal and type *git clone https://github.com/h0tPlug1n/VulnerableApps* this command will clone the repo in your local directory. I might update the repo and add more vulnerable applications for you to test. Feel free to perform a test on those given apps.

But wait, before you jump to the next chapter, which is undoubtedly going to be very interesting, I would like to draw your attention into the brief summary of the tools that we have downloaded and installed now and we'll be using in the future because after all we aren't targeting to become a script kiddie.

Android Studio and its emulator: The official Integrated Development Environment (IDE) for creating Android apps is called Android Studio. For testing and debugging mobile applications, it comes with an emulator (AVD) that enables developers and penetration testers to build virtual Android smartphones.

Burp Suite: Burp Suite is a well-liked tool for evaluating the security of web applications, and it also has functionality for mobile penetration testing. It aids testers in performing many forms of attacks, data transmission and encryption vulnerability analysis, and intercepting mobile application traffic.

Mobile Security Framework: This is an open-source framework for assessing the security of mobile applications that work with both Android and iOS. To find weaknesses, possible risks, and to carry out thorough testing of mobile applications, it combines static and dynamic analysis, decompilation, and malware analysis tools.

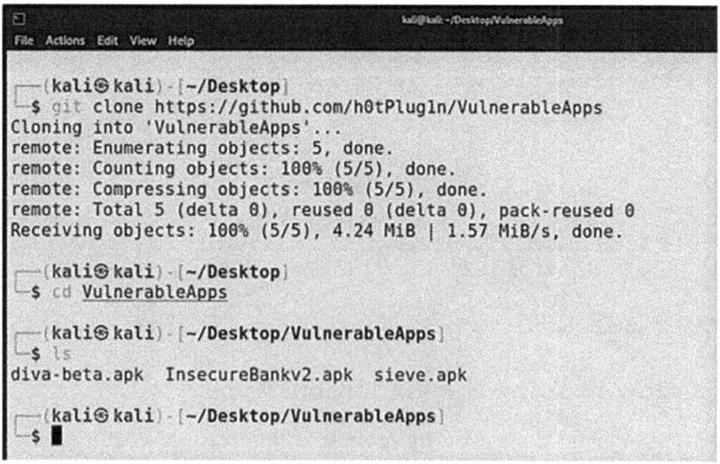

```
┌──(kali㉿kali)-[~/Desktop]
└─$ git clone https://github.com/h0tPlug1n/VulnerableApps
Cloning into 'VulnerableApps'...
remote: Enumerating objects: 5, done.
remote: Counting objects: 100% (5/5), done.
remote: Compressing objects: 100% (5/5), done.
remote: Total 5 (delta 0), reused 0 (delta 0), pack-reused 0
Receiving objects: 100% (5/5), 4.24 MiB | 1.57 MiB/s, done.

┌──(kali㉿kali)-[~/Desktop]
└─$ cd VulnerableApps

┌──(kali㉿kali)-[~/Desktop/VulnerableApps]
└─$ ls
diva-beta.apk  InsecureBankv2.apk  sieve.apk

┌──(kali㉿kali)-[~/Desktop/VulnerableApps]
└─$ ▮
```

Figure 2.119 Downloading the vulnerable apps.

Jadx: It is an open-source decompiler created specifically for examining and reverse-engineering Android applications. Decompiling and analyzing APK files enable penetration testers to comprehend the inner workings of the application and spot any potential security flaws or strange code.

keytool: The JDK contains the command-line tool keytool. In Java-based applications, especially Android apps, it is largely used to manage cryptographic keys, certificates, and keystores. In mobile penetration testing, keytool is helpful for managing activities including SSL/TLS certificates, code signing, and key management.

Drozer: It is a dynamic analysis-capable framework for Android security testing. It enables penetration testers to communicate with Android applications, carry out runtime analysis, find vulnerabilities, create attack simulations, and examine the behavior of the program.

Frida: It is a dynamic instrumentation toolkit with which security researchers can insert custom code into running mobile applications. It facilitates dynamic analysis and testing of mobile applications by enabling runtime manipulation, hooking, and tracing of function calls.

In mobile penetration testing, these tools and programs can be used to build virtual devices, examine traffic, spot vulnerabilities, decompile programs, handle cryptographic keys, carry out static and dynamic analysis, and manipulate runtime. Penetration testers frequently utilize these tools, each of which has unique capabilities and applications, to evaluate the security of mobile devices and applications.

Now, let's jump to the next chapter to learn more about the adb.

Chapter 3

OWASP Mobile Top 10

If you are new to cybersecurity and this is your first experience with the pentesting domain after learning the fundamentals, it's possible that you are unaware of the OWASP community. Therefore, the Open Web Application Security Project, popularly known as OWASP, is a foundation that provides security analyses and recommendations. A nonprofit organization that aims to increase the safety of software. The OWASP Foundation serves as a resource for developers and technologists to protect the web through community-led open-source software projects, hundreds of local chapters globally, tens of thousands of members, and premier educational and training conferences. The OWASP Top 10 list is used by many industries and organizations to rank security issues. The Top 10 Vulnerabilities for Web, Mobile, and APIs are all provided by the OWASP community. Let's have a look at the Top 10 Vulnerabilities for Mobile Applications according to OWASP.

3.1 OWASP MOBILE TOP 10 (2014)

M1: Weak Server-Side Controls: Weak server-side controls, as the term implies, do not communicate data from the mobile application to the server side in a safe and secure manner or expose certain critical APIs when delivering the data. Consider an Android application that authenticates users without verifying their inputs by using their login credentials to the server. An attacker might alter the credentials to get access to private or restricted portions of the system. This weakness might be seen as a weakness in both online apps and mobile applications.

M2: Insecure Data Storage: Simply said, this refers to placing application-related data on the device in a user-accessible location. Secret user or app information is frequently stored by Android apps in shared preferences, SQLite (in plain form), or external storage. Even if an application is keeping sensitive data in the data folders (/data/data/package-name), developers should always bear in mind that once

 DOI: 10.1201/9781003450474-3

the phone is rooted, a malicious application or an attacker will have access to it.

M3: Insufficient Transport Layer Protection: Many Android developers use unsafe methods to transmit data across networks, such as HTTP or improper SSL implementation. As a result, the app is susceptible to all forms of network assaults, including traffic interception, parameter manipulation during data transmission from the application to the server, and response modification to unlock the program's closed sections.

M4: Unintended Data Leakage: Applications become vulnerable to this when they store data in a place that is also vulnerable. The clipboard, URL caches, browser cookies, HTML5 data storage, analytics data, and so forth are a few examples. An illustration would be the Android logs, which are mostly utilized by developers for debugging. The best location to look for leaked data is in the logs. In these situations, adb logcat can assist us in retrieving the Android logs.

M5: Poor Authorization and Authentication: Applications for Android are prone to security flaws if they attempt to verify or approve a user based only on client-side checks. It should be emphasized that after the phone is rooted, the majority of client-side security measures may be disregarded by an attacker. As a result, it is advised that application developers should utilize server-side authentication and authorization with appropriate checks, and probably to use a randomly generated token to authenticate the user on the mobile device.

M6: Broken Cryptography: Simply said, this refers to encrypting the data components using insecure cryptographic techniques. This can include some of the well-known weak points, such MD5, SHA1, or even a specially built one with inadequate security safeguards.

M7: Client-Side Injection: Simply described, it's an attack strategy where the attacker injects a malicious code into the mobile app through input fields, and the app blindly trusts and runs it without sanitizing the input. Since Android applications utilize SQLite3, SQL injection is a frequent example. Since websites and online apps use JavaScript, bugs like XSS would also work if the program was created using WebView.

M8: Security Decisions via Untrusted Input: Inputs like cookies, environment variables, and hidden form fields could be taken for granted by developers as being unchangeable. A hacker might, however, alter these inputs using personalized clients or other assaults. Attackers can get around the software's security even though this change might not be noticed since security choices like authentication and authorization are reliant on the values of these inputs. An example would be a security mechanism used by an application that depends on the existence or values of an input, but the input can be changed by a malicious actor in a way that gets around the protection mechanism.

M9: Improper Session Handling: When handling sessions for a mobile app, the developer must be mindful of a variety of issues, including appropriate authentication cookie expiration, secure token generation, and failure to invalidate sessions, at the backend. Between the online application and the Android application, good, secure synchronization must be maintained.

M10: Lack of Binary Protection: This results in an inability to effectively oppose reversing or decompiling the application. If good developing procedures, which we will cover in this chapter, were not followed, tools like Apktool and dex2jar might be used to reverse an Android application, exposing the application to numerous types of security concerns. There are numerous tools available that could aid the developer in preventing this cause.

3.2 OWASP MOBILE TOP 10 (2016)

M1: Improper Platform Usage: When developers disregard the security safeguards offered by the mobile platform, a vulnerability results. Here are a few typical instances:

Developers may store passwords and other sensitive data in plain text, leaving it open to hacking, rather than using the platform's secure storage features. Sandboxing apps are common practice on mobile platforms to limit their access to system resources. Malicious apps could be able to escape their sandbox and access unapproved data if this functionality is not used. Applications may ask for inappropriate permissions or fail to properly implement them. Attackers may be able to access sensitive information or features as a result of this.

M2: Insecure Data Storage: This category draws attention to the dangers of storing private communications, financial information, and user passwords on a device without proper security measures. Sensitive information, for instance, should always be encrypted both in transit (during communication) and at rest (on the device). If it's stored in an unencrypted manner, attackers can readily access it. It is relatively easy to crack old or weak algorithms, even if the data is encrypted. Decryption requires encryption keys. Negligent key management techniques, such as keeping them on the device, can jeopardize the encryption process as a whole.

M3: Insecure Communication: The flaws in the data transmission between the backend servers and the mobile application are referred to as this vulnerability. Common problems include the ability for attackers to intercept data sent across unencrypted channels, such as plain HTTP, including login passwords. Attackers can get around obsolete or unreliable encryption methods, such as SSL/TLS, by using weak ciphers.

Through communication interceptions between the app and the server, attackers might possibly steal data or insert a malicious code in these attacks.

M4: Insecure Authentication: This category is devoted to flaws in user login procedures that could be used to obtain unauthorized access. Here are a few examples: imposing minimum length constraints on passwords without performing complexity checks makes it simpler for attackers to use brute-force techniques to crack passwords. By requiring a second factor in addition to a password, MFA increases security and lowers the likelihood of illegal access. As was previously indicated, if hackers manage to access the application's data storage, passwords stored in the plain text are vulnerable to breaches.

M5: Insufficient Cryptography: Using antiquated or inadequate encryption techniques to secure data is the source of this issue. Here's how it might be abused: encryption methods that were once thought to be secure, like MD5 or earlier SHA versions, are now deemed to be rather easily cracked. As was previously mentioned, poor key management procedures make encryption useless. Robust random numbers are essential to cryptographic systems. Using weak or predictable random number generators can lead to the encryption process being compromised.

M6: Insecure Authorization: Inadequate controls over the actions users can take within the application are the cause of this vulnerability. These are some typical problems: for example, hackers may use app vulnerabilities to escalate their privileges and access data or capabilities that are not authorized. Inadequate access control measures may provide unauthorized users access to information or features intended for particular user roles. Apps may gather features over time that are no longer needed but still have access to private data. These underutilized qualities could be points of attack.

M7: Client Code Quality: This category emphasizes the importance of writing secure, maintainable, and well-structured client-side code. Poor code quality can introduce vulnerabilities that are often exploited by attackers. Common issues include improper validation and sanitization of user input, insecure error handling, hardcoded credentials, and flawed logic. These mistakes can lead to injection attacks (e.g., XSS, SQLi), data leakage, and other security breaches. Ensuring secure coding practices, such as input validation, secure API usage, and adherence to coding standards, is vital to reduce the attack surface of the mobile application.

M8: Code Tampering: Code tampering refers to the unauthorized modification of a mobile application's code after it has been deployed. Attackers often reverse engineer the app, alter its behavior (e.g., bypassing license checks or authentication mechanisms), and repackage it for malicious use. This category highlights the need for runtime integrity checks, anti-debugging techniques, obfuscation, and tamper detection mechanisms to protect mobile apps from being altered and redistributed in compromised forms. A tampered app can lead to data theft, fraudulent transactions, or user exploitation.

M9: Reverse Engineering: To exploit vulnerabilities, attackers must comprehend the inner workings of the application. How it might provide a security concern is as follows: determining vulnerable vulnerabilities using network traffic analysis or app code analysis. Attackers may find holes in security that they can exploit to obtain control or access without authorization. Once vulnerabilities are found, hackers can design unique exploits to take advantage of those flaws and breach the application.

M10: Extraneous Functionality: This category deals with extraneous features or the code that may enhance the app's attack surfaces. This is how it might cause issues. An unused code adds to the overall size and complexity of the application, making it more difficult to maintain and perhaps harboring vulnerabilities. Using out-of-date or unsafe third-party libraries can make the application more vulnerable.

3.3 OWASP MOBILE TOP 10 (2024)

M1: Improper Platform Usage: Apps' methods for storing, transmitting, and validating login credentials are vulnerable. This can be used by attackers to obtain unauthorized access or steal passwords. This can entail sending credentials across unencrypted connections, enabling lax password restrictions, or storing passwords in an unsecured manner.

M2: Inadequate Supply Chain Security: Attackers may take advantage of flaws in outdated dependencies or third-party libraries that are embedded into the application. They may be able to access the main program through this vulnerability and take user data.

M3: Insecure Authentication/Authorization: It is simpler for attackers to get around security and access accounts or features within the app without authorization when there are lax authentication measures in place, such as unencrypted password storage or no multifactor authentication.

M4: Insufficient Input/Output Validation: Attackers can insert a harmful code into apps when they fail to adequately check user inputs (such as data entered into forms) or outputs (such as data transmitted to servers). This code has the ability to propagate malware, take over systems, and steal data. SQL injection and Cross-Site Scripting (XSS) assaults are two examples.

M5: Insecure Communication: Sensitive information can be intercepted by hackers if it is sent over unencrypted networks, such as plain HTTP, and includes passwords, financial information, or personal details. By utilizing HTTPS, which encrypts communication between the app and the server, this can be lessened.

M6: Inadequate Supply Chain Security: Apps that gather a lot of data or store it in an unencrypted manner may endanger users' privacy. This may entail gathering more information than is required for the program to function or improperly anonymizing user data.

M7: Insufficient Binary Protection: Mobile apps are insecure if they don't have built-in safeguards against hacking or reverse engineering. Attackers can use these flaws to gain insight into the inner workings of the app and possibly identify further vulnerabilities. One can use methods such as code obfuscation to increase the difficulty of reverse engineering.

M8: Security Misconfiguration: Vulnerabilities may become visible in the app or development environment due to incorrect security settings. Errors during app deployment or insecure default configurations may be to blame for this.

M9: Insecure Data Storage: Encrypting sensitive data, such as bank information or passwords, leaves the device open to theft in the event that malware compromises the device or it is physically stolen. Sensitive data must be encrypted when it is at rest.

M10: Insufficient Cryptography: Data is susceptible to decryption by attackers when inadequate encryption algorithms are used or when encryption is not used at all. This may entail transferring or storing data without any encryption at all, or it may entail employing antiquated encryption standards.

Security is a multilayered strategy rather than a single fix. While addressing the risks listed in the OWASP Top 10 is important, it is only one aspect of the puzzle. Mobile security can be further improved with additional steps like educating users on secure password management and safe app usage practices.

Through comprehension of these hazards and execution of an all-encompassing security plan, developers can construct mobile applications that are not just operational but also reliable and safe for end users.

Chapter 4

Introduction to adb

You may interface with a device using the flexible command-line tool known as adb. The adb command makes it easier to do a number of device-related tasks, including installing and troubleshooting apps. It also gives you access to a Unix shell, so you may execute various commands on a device.

There are three parts to this client-server program:

- **Client:** A client that transmits commands. On your development computer, the client runs. An adb command can be used to launch a client from a command-line terminal.
- **Daemon (adbd):** A device's daemon (adbd), which executes commands. On each device, the daemon operates as a background process.
- **Server:** The daemon and client's communication are handled by a server. The server operates on your development PC in the background.

There are many commands and many functionalities that adb can perform, but we will focus on the necessary ones:

1. Listing all the attached devices: *adb devices*
2. Listing a long list of all the attached devices: *adb devices -l*
3. Connecting a device: *adb connect device_ip:port*
4. Pushing a file into a device: *adb push <path/to/local_filename> <path/to/remote_filename>*
5. Pulling a file from the device: *adb pull <path/to/remote_filename> <path/to/local_filename>*
6. Installing an app to a connected device: *adb install <path/to/app.apk>*
7. Stop adb server: *adb kill-server*
8. Port forwarding: *adb forward tcp:from_port tcp:to_port*
9. Get shell access: *adb shell*
10. Restart adbd with root permission: *adb root*
11. Restart adbd without root permission: *adb unroot*
12. Issuing shell commands: *adb shell command*

DOI: 10.1201/9781003450474-4

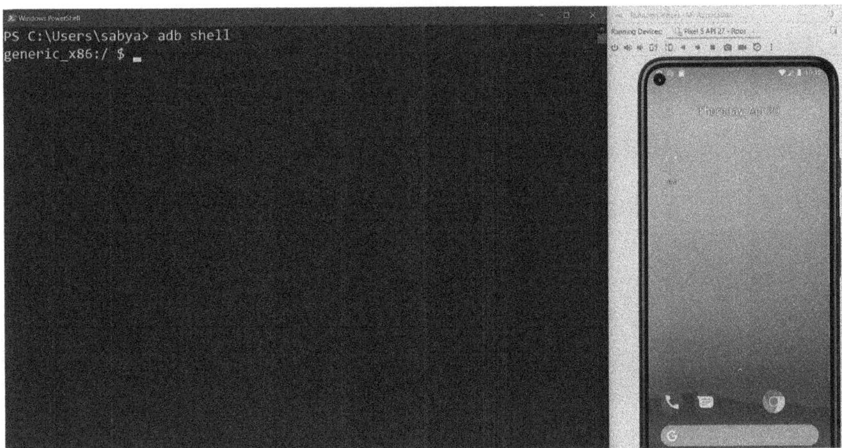

Figure 4.1 Access to the shell of the phone.

```
PS C:\Users\sabya> adb shell
generic_x86:/ $ pwd
/
generic_x86:/ $ whoami
shell
generic_x86:/ $ exit
PS C:\Users\sabya>
```

Figure 4.2 Exiting out of shell.

```
generic_x86:/ $ exit
PS C:\Users\sabya> adb root
restarting adbd as root
PS C:\Users\sabya> adb shell
generic_x86:/ # id
uid=0(root) gid=0(root) groups=0(root),1004(input),1007(log),1011(adb),10
15(sdcard_rw),1028(sdcard_r),3001(net_bt_admin),3002(net_bt),3003(inet),3
006(net_bw_stats),3009(readproc),3011(uhid) context=u:r:su:s0
generic_x86:/ #
```

Figure 4.3 Shell having root privilege.

```
generic_x86:/ # exit
PS C:\Users\sabya> adb unroot
restarting adbd as non root
PS C:\Users\sabya> adb shell
generic_x86:/ $ _
```

Figure 4.4 Shell having non-root privilege.

```
Windows PowerShell
PS C:\Users\sabya> adb push sabyasachi.txt storage/emulated/0/Download
sabyasachi.txt: 1 file pushed, 0 skipped. 0.1 MB/s (40 bytes in 0.001s)
PS C:\Users\sabya> _
```

Figure 4.5 Sending a text file into the phone via adb.

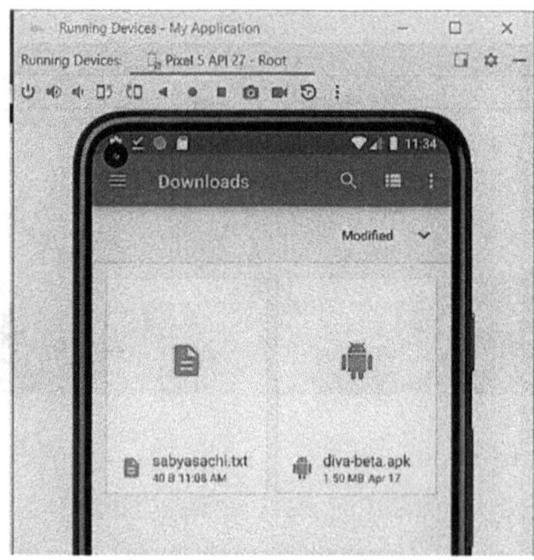

Figure 4.6 File manager of the phone.

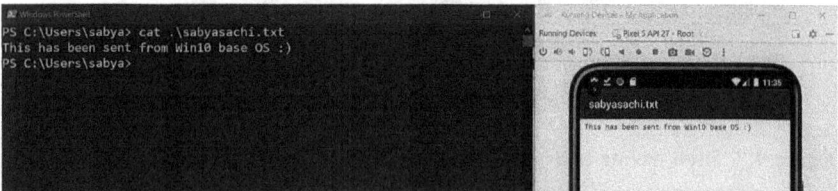

Figure 4.7 The file sabyasachi.txt been sent via adb.

There are other command-line tools like adb, but those are beyond the scope of this book. I might cover those in my other publications. Within the adb you can issue several other commands like am (activity manager), pm (package manager), etc. about which we'll look upon those in Chapter 6.

Now let's see adb in action. For that I am going to start an emulator and then we will observe most of the common functionalities that adb provides.

The prompt *generic_x86/$* represents that we have got access to the shell of the phone. The command used here was – *adb shell*

Now like Linux, we can execute commands like – *ls, pwd, whoami, id*, etc.

Now, the shell access we've got is of normal low privilege user. To gain root access, at first exit out of the shell using – *exit*

Then, type – *adb root* followed by *adb shell* again to get root access.

To get back to unprivileged access, use command – *adb unroot* and then use the command – *adb shell.*

Note: Check whether the prompt is $ or #. For non-root it is $ and # for root.

Now, let's see the use case of the *adb push* and *adb pull* command.

For that, I've created a file named – sabyasachi.txt and going to send it via adb into my phone.

For that, type – *adb push sabyasachi.txt storage/emulated/0/Download.* This command requires root privilege.

Now, whether the file has been received or not can be verified by directly checking the phone.

If we check the integrity of the file, we can see that the exact file has been sent.

Similarly, *adb pull* does the opposite. It pulls the given file into the current path. There are other commands too. Go give it a try and understand.

Chapter 5

Reversing and traffic interception of apps

This chapter will discuss reversing an app, which is a crucial step in static analysis, how to examine the source code of an apk and reverse it to look for any security holes.

In addition, we'll also demonstrate how to intercept and capture incoming and outgoing communication from an Android smartphone. As a pentester, it is your obligation to be able to intercept any signals sent by mobile applications that include sensitive information and safeguard the communication channel.

There are several methods for reversing an Android app, but I'll explain the three simplest ones here with the intention of appealing to beginners. The tools we'll be focusing upon are Jadx-Gui, apktool, and dex2jar. Now, let's go through the practical approach or reversing apps.

5.1 REVERSING APPS USING JADX-GUI

Step #1: Go to the application menu > Reverse Engineering > jadx-gui and click on it.

Step #2: Once jadx loaded, click on the Open file and select your apk from the dialog box appeared.

Step #3: Click OK and after sometime Jadx will present you with all the app folders and files.

In the left pane, all the directories and files that could be present in an app are listed and on the right-hand side is the AndroidManifest.xml file which I clicked to view it.

Now, let's see a second approach to achieving the same.

DOI: 10.1201/9781003450474-5

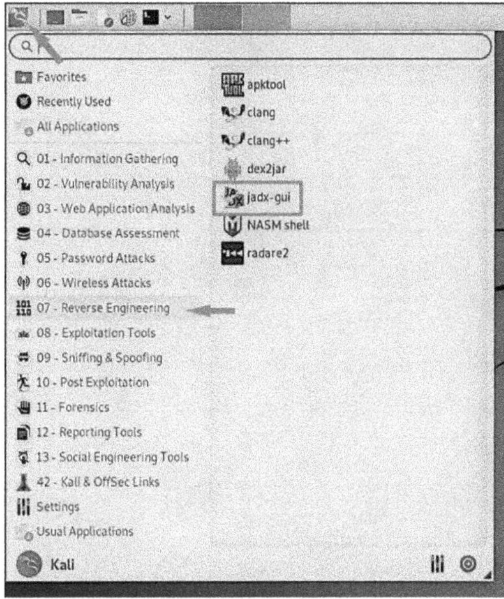

Figure 5.1 Launching jadx-gui.

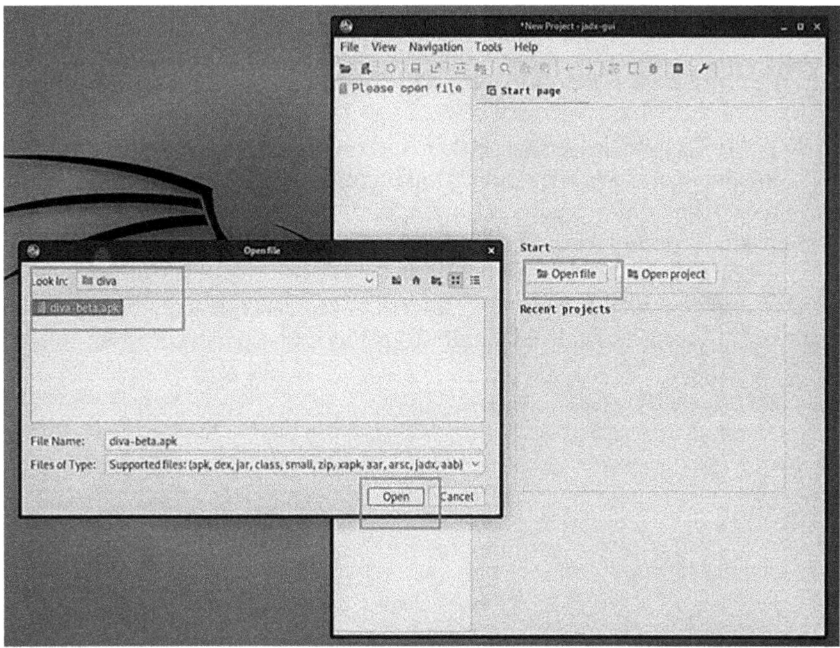

Figure 5.2 jadx-GUI target file selection window.

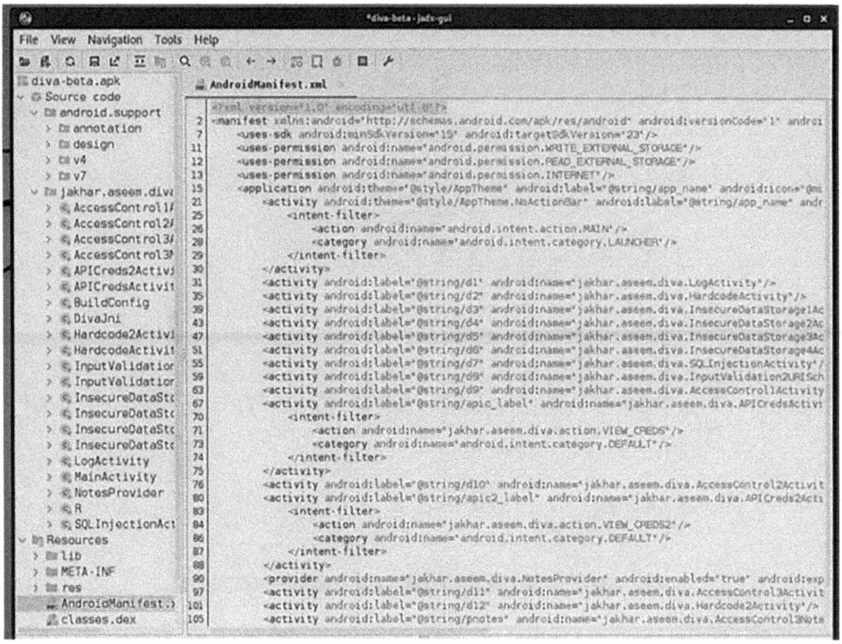

Figure 5.3 Output of jadx-GUI.

5.2 REVERSING APPS USING APKTOOL

Step #1: To launch keytool, either open terminal and type *apktool* then hit Enter or go to the application menu > Reverse Engineering > Apktool. Click on it to launch.

Step #2: Type in terminal *apktool d appname.apk*, where 'd' stands for decompile. Hit enter and it will do its job.

The above highlighted files and folders have been extracted by apktool.

Step #3: Opening AndroidManifest.xml in mousepad and trying to read the code

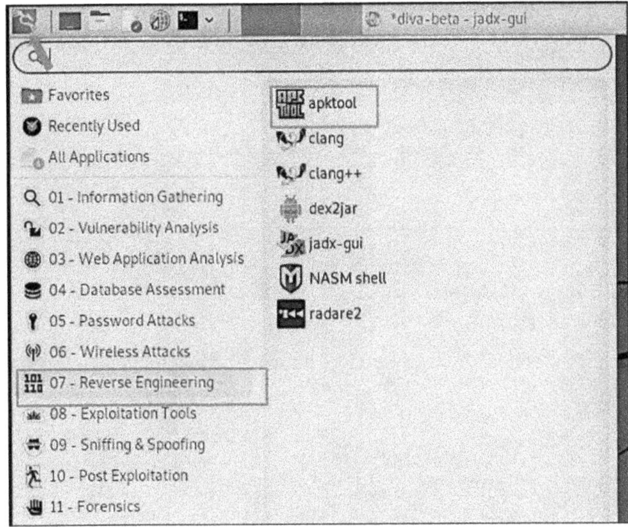

Figure 5.4 Launching apktool via application menu.

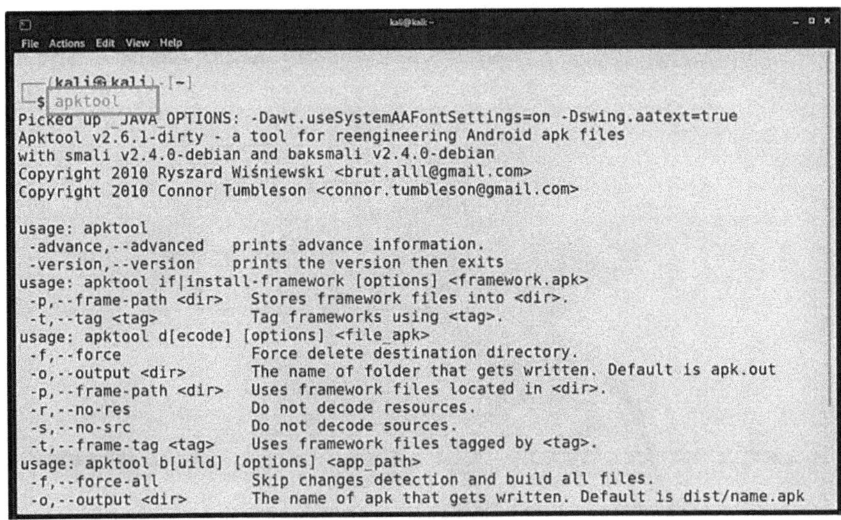

Figure 5.5 Launching apktool via terminal.

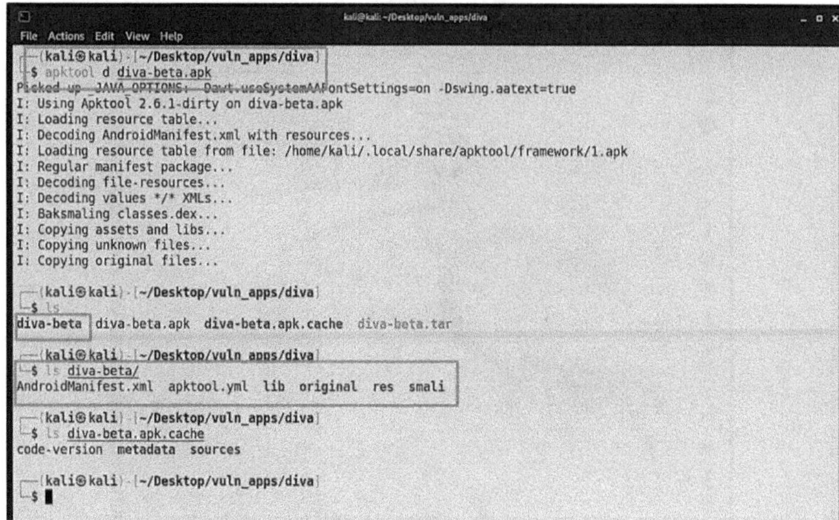

Figure 5.6 The output of apktool.

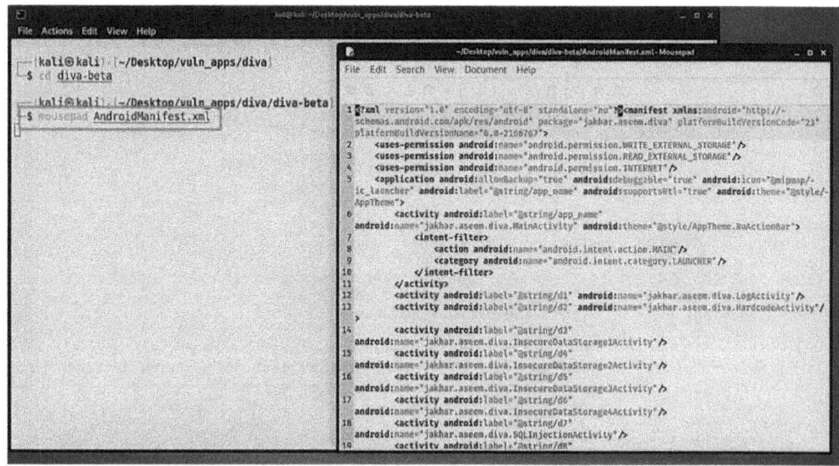

Figure 5.7 AndroidManifest.xml file decompiled by apktool.

5.3 REVERSING APPS USING DEX2JAR

Step #1: Open terminal and type *d2j-dex2jar appname.apk* and hit Enter. This will decompile the app for you as a .jar file. The arrow-marked file in Figure 4.8 is the output.

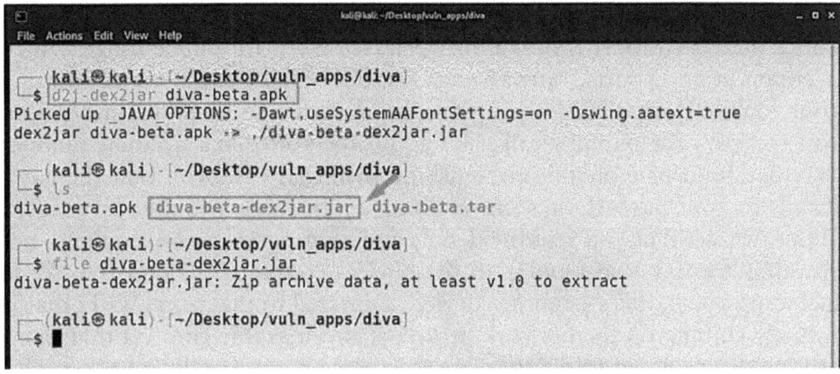

Figure 5.8 dex2jar in action.

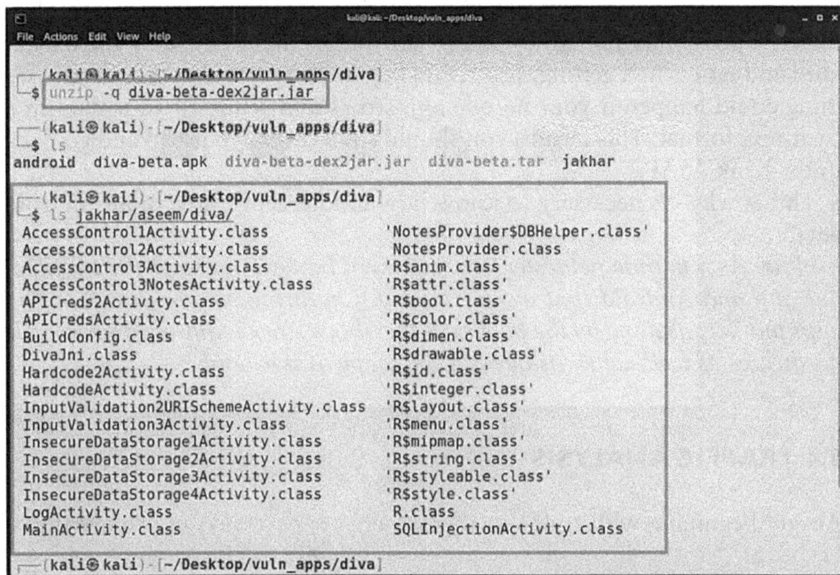

Figure 5.9 The extracted data which can be viewed using any text editor.

Step #2: Extracting the archive to access the source code.

Now since we explored different ways to reverse or decompile an apk file, let's move forward to the excitement part which is to intercept Android traffic. But before beginning, let me give you a scenario to make you explain its necessity.

Consider you are at an airport, waiting for your flight. Meanwhile, you notice that the airport is providing open free Wi-Fi for public usage, which is common at airports, railway stations, bus stands, etc. Now connecting your mobile to that Wi-Fi, you thought of watching a movie. In between you receive your parent's call asking you for your bank account number due to certain unavoidable circumstances. In that case, you sent the bank details to your parents via some chatting application. You are rest assured about the fact that you sent the details and your parents received them, but you didn't know that among all the devices connected to that free Wi-Fi including yours, there's another device connected to that same Wi-Fi that is actually sniffing on the network. It means all the traffic goes via that intercepting device. In addition, that device is being controlled by a hacker. This also states that your data too went to your parent via his device and he can actually see the incoming and outgoing traffic. This time, two things will happen. If your mobile application properly implements security protocol and implements complex encrypting algorithms to encrypt your data, then there's a possibility that your data is safe and cannot be read by the hacker until and unless he's getting the private key to decrypt it. Another disastrous thing could happen if your mobile app stores and sends all your data in a clear-text format. This means, you should soon cancel your trip and go back to the bank ASAP:)

This is why it's necessary to know how to intercept traffic of an Android device.

Note: As a mobile pentester, we will learn how to intercept the traffic of our app and Android that we are testing. Capturing network traffic of all apps publicly as done by the hacker in the above story requires some specific hardwares as well, which is beyond the scope of this book.

5.4 TRAFFIC ANALYSIS

Alright beginning with traffic analysis, there are two ways to achieve this:

1. **Active analysis:** Actively snooping on every network request the program makes and immediately modifying it. Here, a proxy will be set up, and all network calls made and received will go via it.
2. **Passive analysis:** There is no active network communication interception. Instead, we just succeed in capturing the network packets, which we then open using some packet analysis tools to inspect the vulnerability and security issues that are present.

First and foremost, you must install adb on your host system, a Windows computer in this case, before doing any analysis or pentesting. You'll discover the cause afterward. Therefore, put "download android-platform-tools" into Google. You will receive a link to download it from the official

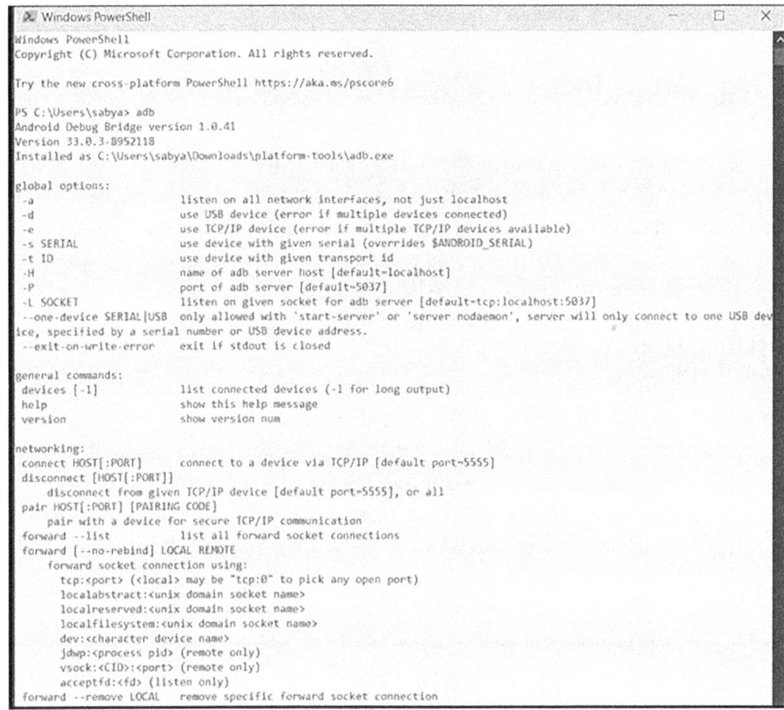

Windows PowerShell
Windows PowerShell
Copyright (C) Microsoft Corporation. All rights reserved.
Try the new cross-platform PowerShell https://aka.ms/pscore6
PS C:\Users\sabya> adb
Android Debug Bridge version 1.0.41
Version 33.0.3-8952118
Installed as C:\Users\sabya\Downloads\platform-tools\adb.exe
global options:
-a listen on all network interfaces, not just localhost
-d use USB device (error if multiple devices connected)
-e use TCP/IP device (error if multiple TCP/IP devices available)
-s SERIAL use device with given serial (overrides $ANDROID_SERIAL)
-t ID use device with given transport id
-H name of adb server host [default=localhost]
-P port of adb server [default=5037]
-L SOCKET listen on given socket for adb server [default=tcp:localhost:5037]
--one-device SERIAL|USB only allowed with 'start-server' or 'server nodaemon', server will only connect to one USB dev
ice, specified by a serial number or USB device address.
--exit-on-write-error exit if stdout is closed
general commands:
devices [-l] list connected devices (-l for long output)
help show this help message
version show version num
networking:
connect HOST[:PORT] connect to a device via TCP/IP [default port=5555]
disconnect [HOST[:PORT]]
disconnect from given TCP/IP device [default port=5555], or all
pair HOST[:PORT] [PAIRING CODE]
pair with a device for secure TCP/IP communication
forward --list list all forward socket connections
forward [--no-rebind] LOCAL REMOTE
forward socket connection using:
tcp:<port> (<local> may be "tcp:0" to pick any open port)
localabstract:<unix domain socket name>
localreserved:<unix domain socket name>
localfilesystem:<unix domain socket name>
dev:<character device name>
jdwp:<process pid> (remote only)
vsock:<CID>:<port> (remote only)
acceptfd:<fd> (listen only)
forward --remove LOCAL remove specific forward socket connection

Figure 5.10 adb running in PowerShell.

developer. Android website. The .zip archive may be downloaded by clicking that link. Go inside the folder after extracting the zip file. Copy the whole path of the folder and add it into your System environment variables under Path variable. Write *adb* in PowerShell right away. This is what should be visible to you: Figure 5.10. You're good to go if you replied "Yes."

Download link (android-platform-tools) – click here

Now that the emulator is running on our host system and the majority of our pentesting will be done from our Kali VM, and it becomes fairly challenging to create a connection between our VM and emulator using adb. Hence, follow these instructions to connect to the emulator using adb within a virtual machine. Also, this is why there was a need to install adb in Windows as well.

Type in your Host machine (in PowerShell):

1. taskkill /f /t /im adb.exe
2. adb -a nodaemon server
3. Accept the Firewall UAC (User Account Control) prompt
4. Minimize the screen and copy the IP of your host machine. We'll need it.

Type in your Kali VM (in Terminal):

5. adb -H <ip_of_host_machine> -P 5037 shell

Note: While making a connection between your emulator and VM, for some readers it might not work. Make sure you are on a /24 network. That means your Host machine's IP should range between 192.0.0.0 and 223.255.255.255 and your subnet mask is 255.255.255.0. Use *ipconfig* in Windows to verify the addresses. Also, make sure that the VM is bridged as the VM and the host machine should be on the same network.

It should look like this.

Also, sometimes step 2 might not even work. In that case, try changing the port by issuing the command – adb -P <any random port> -a nodaemon server and then in the Kali Machine issue the command – adb -P <the port number> -H <host machine IP> shell. This would solve the problem.

Now, it's expected that you are understanding the commands, but if you don't, no worries we've got you covered. Chill! You're in safe hands:) But

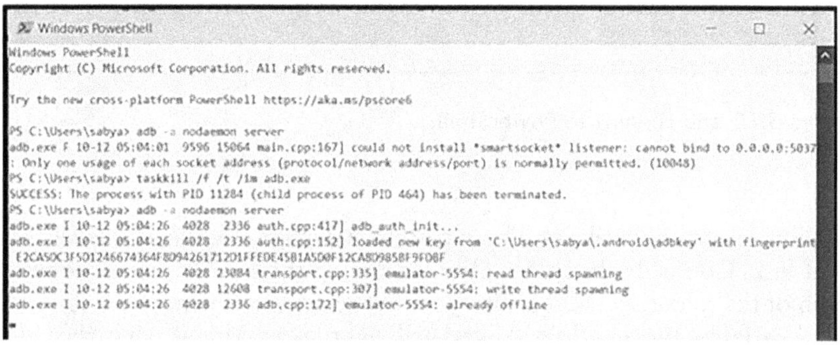

```
Windows PowerShell
Windows PowerShell
Copyright (C) Microsoft Corporation. All rights reserved.

Try the new cross-platform PowerShell https://aka.ms/pscore6

PS C:\Users\sabya> adb -a nodaemon server
adb.exe F 10-12 05:04:01  9596 15064 main.cpp:167] could not install *smartsocket* listener: cannot bind to 0.0.0.0:5037
: Only one usage of each socket address (protocol/network address/port) is normally permitted. (10048)
PS C:\Users\sabya> taskkill /f /t /im adb.exe
SUCCESS: The process with PID 11284 (child process of PID 464) has been terminated.
PS C:\Users\sabya> adb -a nodaemon server
adb.exe I 10-12 05:04:26  4028  2336 auth.cpp:417] adb_auth_init...
adb.exe I 10-12 05:04:26  4028  2336 auth.cpp:152] loaded new key from 'C:\Users\sabya\.android\adbkey' with fingerprint
 E2CA5OC3F5D12466743 64F8D94261712D1FFEDE4581A5D0F12CA8D9858F9FD8F
adb.exe I 10-12 05:04:26  4028 23084 transport.cpp:335] emulator-5554: read thread spawning
adb.exe I 10-12 05:04:26  4028 12608 transport.cpp:307] emulator-5554: write thread spawning
adb.exe I 10-12 05:04:26  4028  2336 adb.cpp:172] emulator-5554: already offline
```

Figure 5.11 Windows PowerShell.

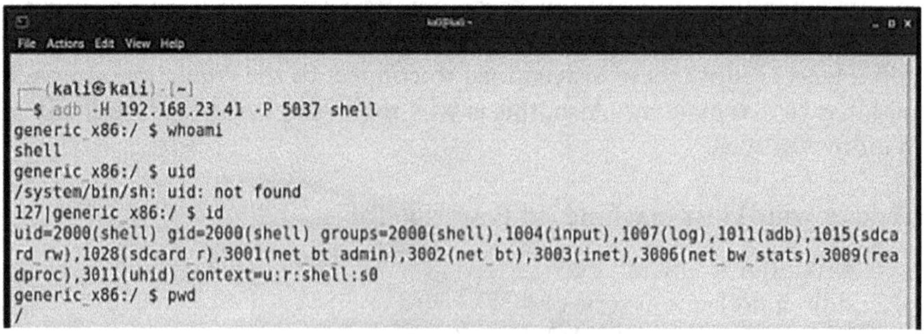

```
File Actions Edit View Help

┌──(kali㉿kali)-[~]
└─$ adb -H 192.168.23.41 -P 5037 shell
generic_x86:/ $ whoami
shell
generic_x86:/ $ uid
/system/bin/sh: uid: not found
127|generic_x86:/ $ id
uid=2000(shell) gid=2000(shell) groups=2000(shell),1004(input),1007(log),1011(adb),1015(sdca
rd_rw),1028(sdcard_r),3001(net_bt_admin),3002(net_bt),3003(inet),3006(net_bw_stats),3009(rea
dproc),3011(uhid) context=u:r:shell:s0
generic_x86:/ $ pwd
/
```

Figure 5.12 Kali terminal having access to the shell of the Android emulator.

now that's not important. What's necessary is to focus upon traffic analysis. Now, since the environment is set up, let's move ahead.

5.5 ANALYSIS OF ANDROID APPS

For the passive analysis, let's quickly download a tool called tcpdump in our Kali box. tcpdump comes already installed in Kali Linux, but we are downloading the file compiled for android.

Step #1: Open Browser in the VM. It could be Firefox, Chromium, or anything and search for "tcpdump for mobile" in the web. Go to this website.

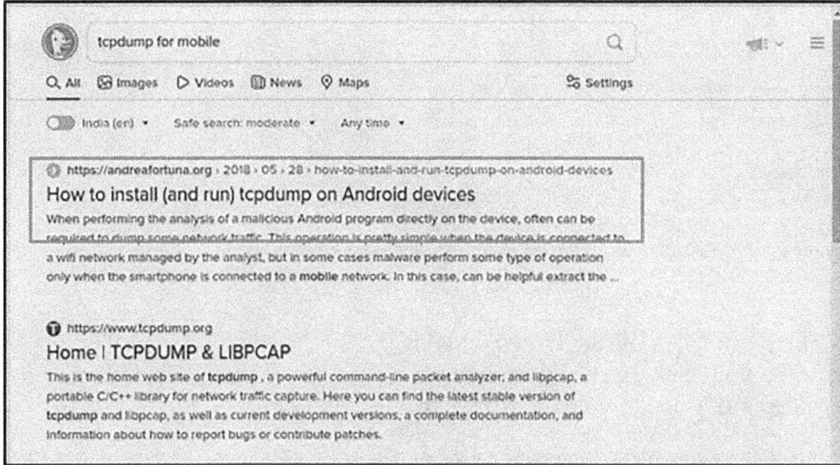

Figure 5.13 Searching for tcpdump in Internet.

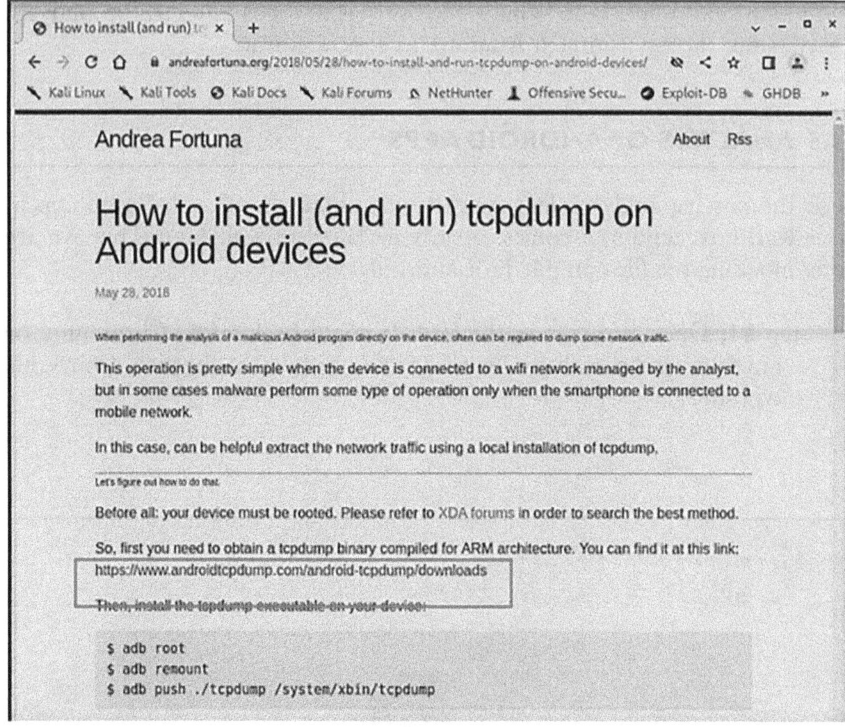

Figure 5.14 Official website of mobile version of tcpdump.

Step #2: Click on the Downloads page.
Step #3: Click on the latest release.
Step #4: Your download will begin.

tcpdump is a command-line utility, or in a simple term, it is a tool that is used for capturing packets and performing analysis over it. We are downloading this tool to capture the Android traffic so that later we can perform the analysis.

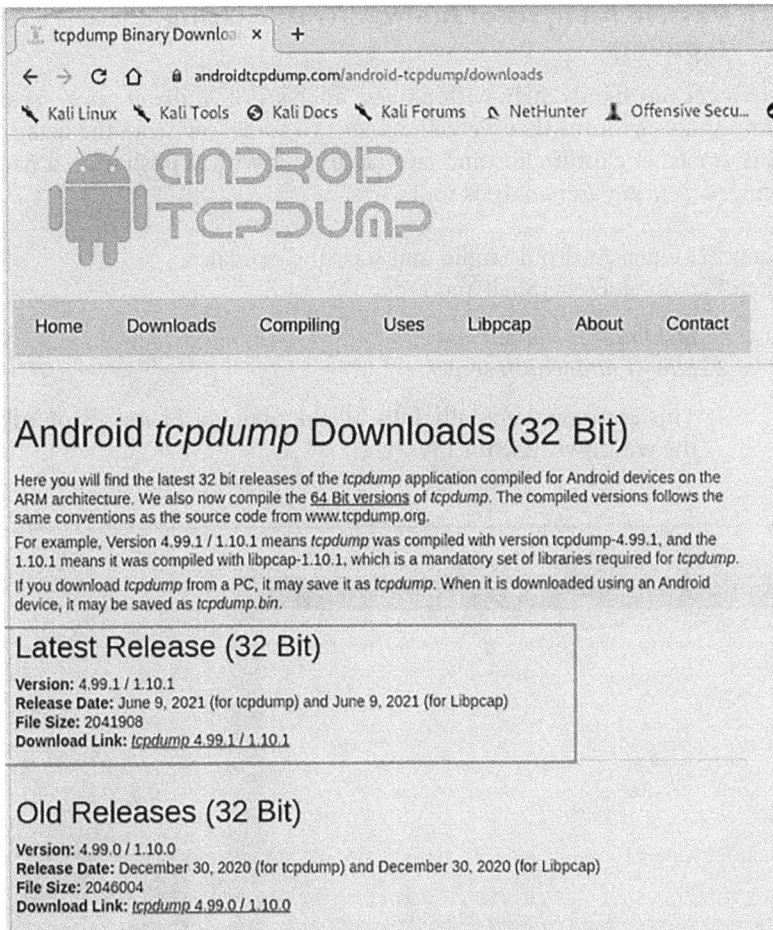

Figure 5.15 Download page of tcpdump.

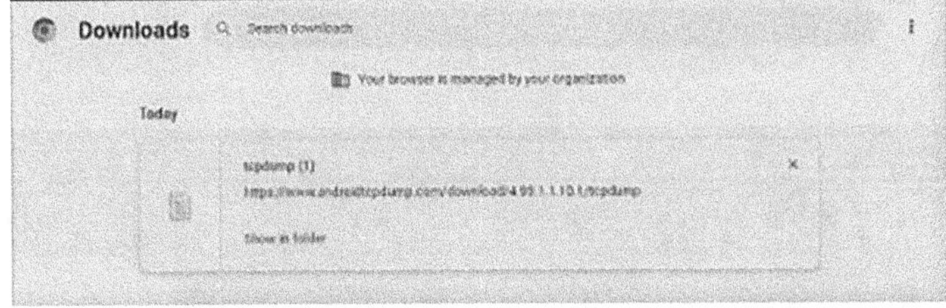

Figure 5.16 The download prompt.

5.5.1 Passive Analysis of Android Traffic Using tcpdump

So, let's get our hands dirty with passive analysis of Android apps. As said, passive analysis means that we will manage to somehow store the .pcap file, that is, a packet capture file, and later analyze it with Wireshark – a packet sniffing, capturing and analysis tool.

Step #1: Open Android Studio and start the emulator.
Step #2: Go to PowerShell and type:

1. *taskkill /f /t /im adb.exe*
2. *adb -a nodaemon server*

1. This command actually kills all the ongoing processes of adb in the windows machine.

Figure 5.17 Kali VM on the left and emulator on the right.

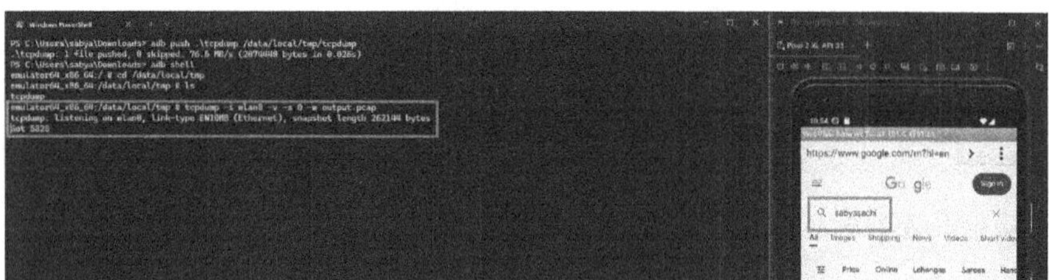

Figure 5.18 PowerShell commands.

```
emulator64_x86_64:/data/local/tmp # tcpdump -i wlan0 -v -s 0 -w output.pcap
tcpdump: listening on wlan0, link-type EN10MB (Ethernet), snapshot length 262144 bytes
^C5830 packets captured
5830 packets received by filter
0 packets dropped by kernel
emulator64_x86_64:/data/local/tmp # ls
output.pcap  tcpdump
emulator64_x86_64:/data/local/tmp # exit
PS C:\Users\sabya\Downloads> adb pull /data/local/tmp/output.pcap .
/data/local/tmp/output.pcap: 1 file pulled, 0 skipped. 73.8 MB/s (6469696 bytes in 0.084s)
```

Figure 5.19 Connecting to the emulator via adb in VM.

Figure 5.20 Changing permission of tcpdump binary file.

2. After the task is killed, it restarts to listen on all interfaces including our VM.

Step #3: Go to Kali VM and push the tcpdump file to the Android emulator by typing: *adb -H <windows_ip> -P 5037 push tcpdump /data/local/tmp/tcpdump*.

1. Location of the tcpdump file present in our machine
2. Location in the emulator where the tcpdump binary file will be kept

So, the command goes like – we are telling adb to push a file or send a file named tcpdump present in our Downloads folder to the Android's /data/local/tmp folder as tcpdump, because that is the place where we will be getting no permission issues. -H is host ip, that is, your window's ip because the emulator is running on Windows and the -P 5037 is the port where the adb connects.

Step #4: Now, gain a root shell to the emulator to change the permission of the tcpdump file and make it executable.

1. Gaining the shell using adb, but logged in as non-root user (check the $ prompt below)
2. Restarted adbd as root permission

3. Gained the shell and now logged as root user (check the # prompt below)
4. Changing permission to read, write, and execute the file

If you are getting $ in the terminal, that means you are logged in non-root. You can type id to confirm your root privilege.

Step#5: Executing tcpdump in the shell of the emulator and doing some web operations in the emulator browser.

Therefore, we are executing tcpdump in the shell of the phone, and on the phone itself, we are trying to open a browser and performing a login operation. Now, since it is web traffic, tcpdump should catch it. As shown in Figure 5.20, it captured 66 packets.

Now, looking upon the command we've written it says – tcpdump to listen on interface (-i) wlan0, that is, Wi-Fi, and give us a verbose output, that is, a detailed output about what is happening in the backend. -s stands for Snapshot length, referred to as snaplen. The amount of bytes that are taken in for each packet is equal to snaplen. You might be able to store more packets if your snaplen is less than the network's maximum packet size. In addition, -w is the file needed to write the output.

After we've performed the login operation on the website, we'll press ^C (Ctrl+C) to stop the tcpdump and then type Exit.

Step #6: Exit tcpdump and pull the pcap file.

Now, we need to pull the output.pcap file. So for that we need to type: *adb -H <host_ip> -P 5037 pull /data/local/tmp/output.pcap Downloads/output.pcap*

/data/local/tmp/output.pcap is the location in the emulator where the pcap file is stored, and Downloads/output.pcap is the location in our machine where we want the file to keep.

Step #7: Use wireshark to analyze the packets.

Now, we'll fireup Wireshark – a GUI packet capturing and analysis tool to observe the packets. For that, in the terminal type: *wireshark path/to/output.pcap*

So, if we properly analyze the packets, we can see that packet no. 22 uses the HTTP protocol, which transmits data in a clear-text format. Also, in the info column, the HTTP method used is POST hence, this packet is sending something on the server. This probably could be the username and password.

If we click into it, we can see the complete detail of the packet. On the right-hand side is the web request that has been made.

On the bottom left, we can see the username and password sent to the server. Hence, this is a win-win situation and as a hacker we have retrieved the username and password.

This was the practical demonstration of how passive traffic analysis is done.

5.5.2 Active Analysis of Android Traffic Using Burp Suite

Now, let's look upon the active analysis part. For this, we just need to deal with the windows machine. Simply launch the emulator and Burp Suite and we are good to go.

Step #1: Go to Burp Suite and click on Proxy Tab > Options. Click on the Add button under Proxy Listeners.
Step #2: Give the port 8085 (you can give anything), select All interface, and then click OK.
Step #3: Click Yes to confirm the security change.
Step #4: Click on Import/Export CA Certificate.
Step #5: Click on Certificate in DER Format and click Next.
Step #6: Click on Select File and choose a location to export Burp certificate.
Step #7: Name the certificate whatever you want with a .CER extension.
Step #8: Open the explorer and drag-and-drop the certificate into the emulator.
Step #9: Go to the Settings menu in the Android emulator.
Step #10: Click on Network & Internet.
Step #11: Click on Wi-Fi options.
Step #12: Click on Wi-Fi Preferences.

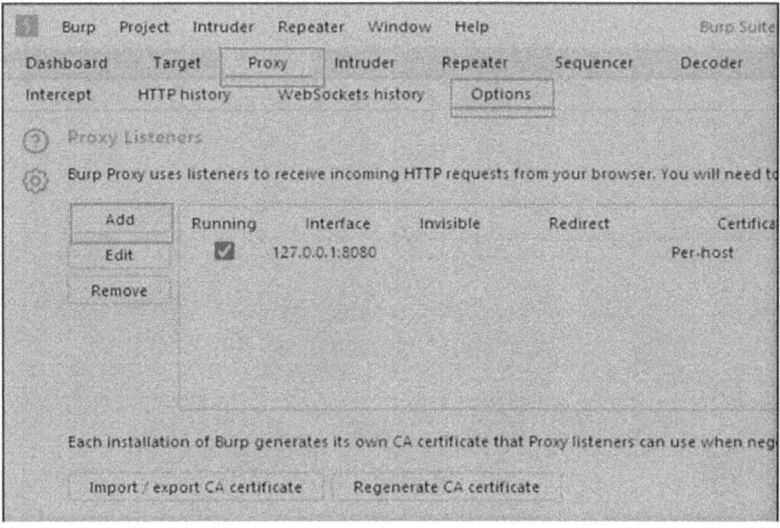

Figure 5.21 Configuring Burp Suite for active analysis.

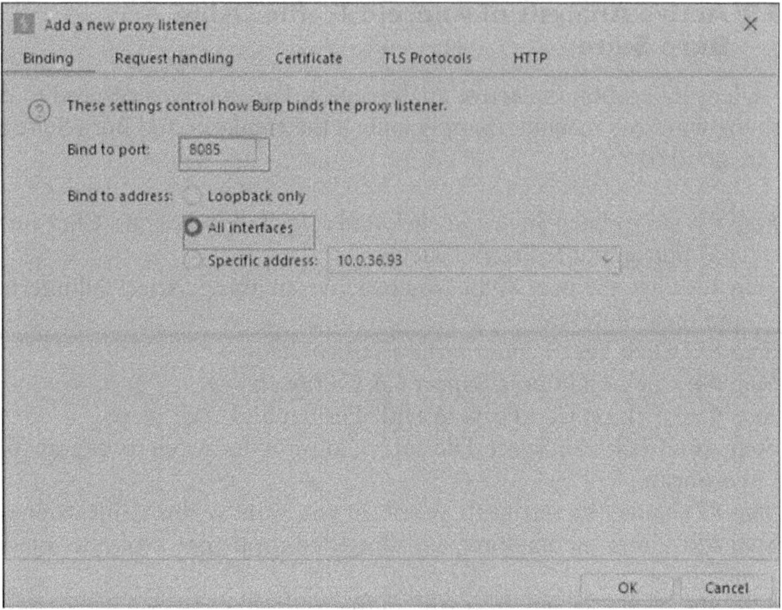

Figure 5.22 Setting up Burp Suite for active analysis.

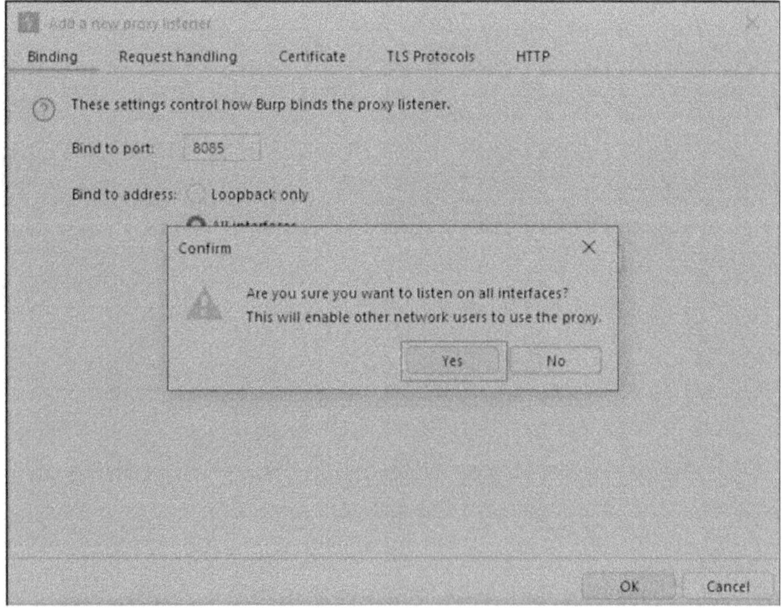

Figure 5.23 Setting up permission.

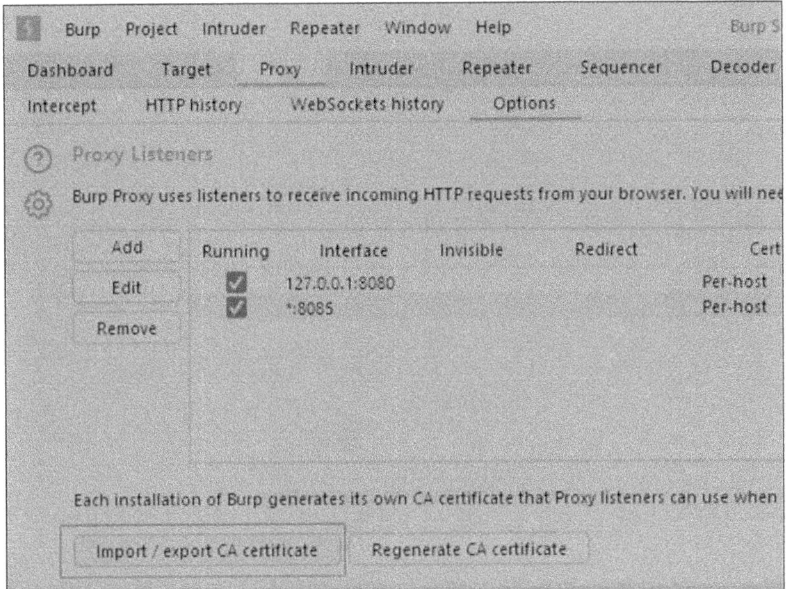

Figure 5.24 Exporting CA Certificate.

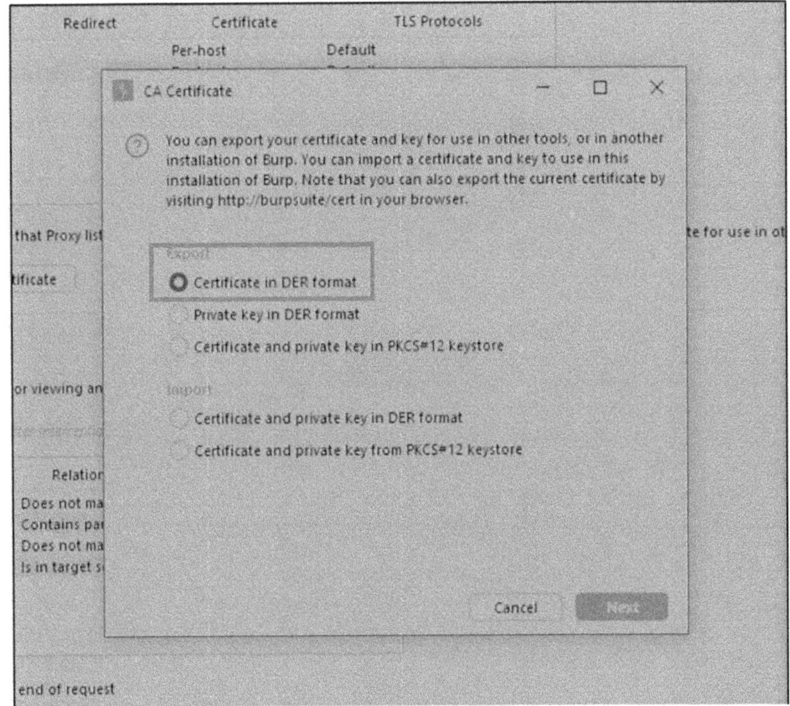

Figure 5.25 Exporting the CA Certificate.

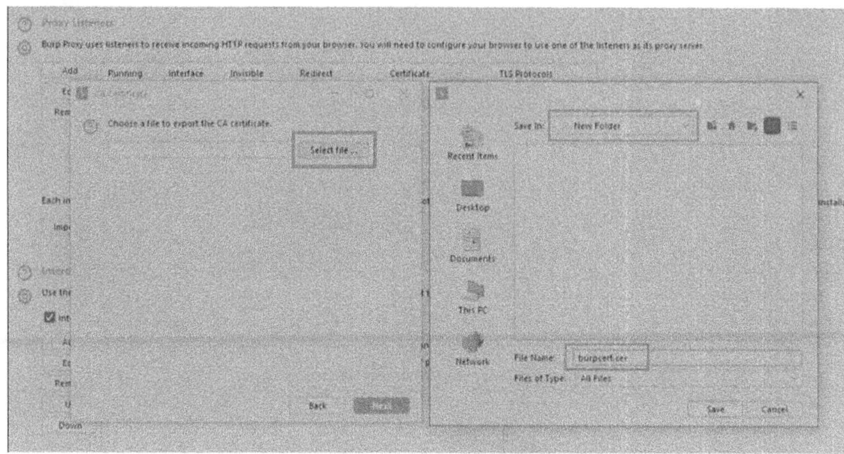

Figure 5.26 Choosing specific file path for exporting.

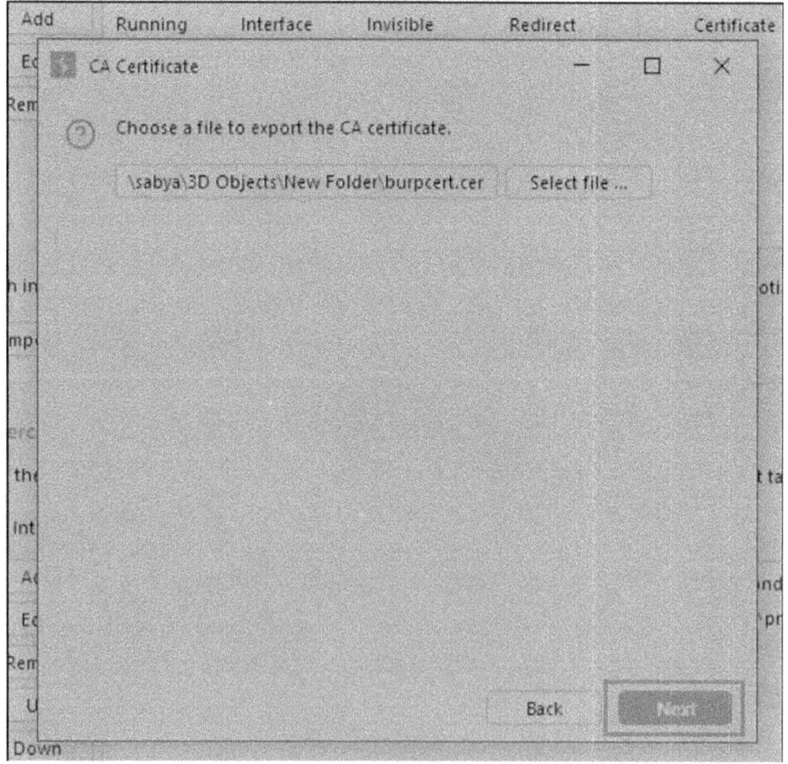

Figure 5.27 Saving the certificate in the selected file path.

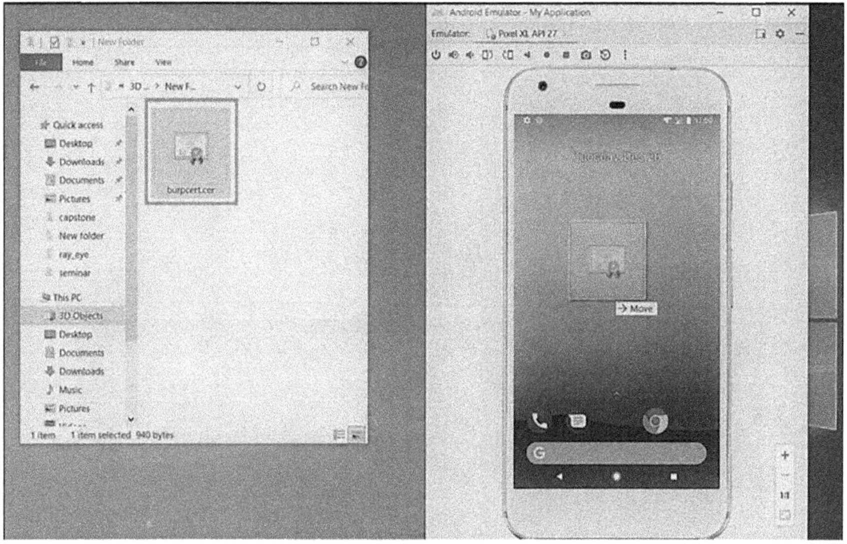

Figure 5.28 Moving the certificate in the emulator.

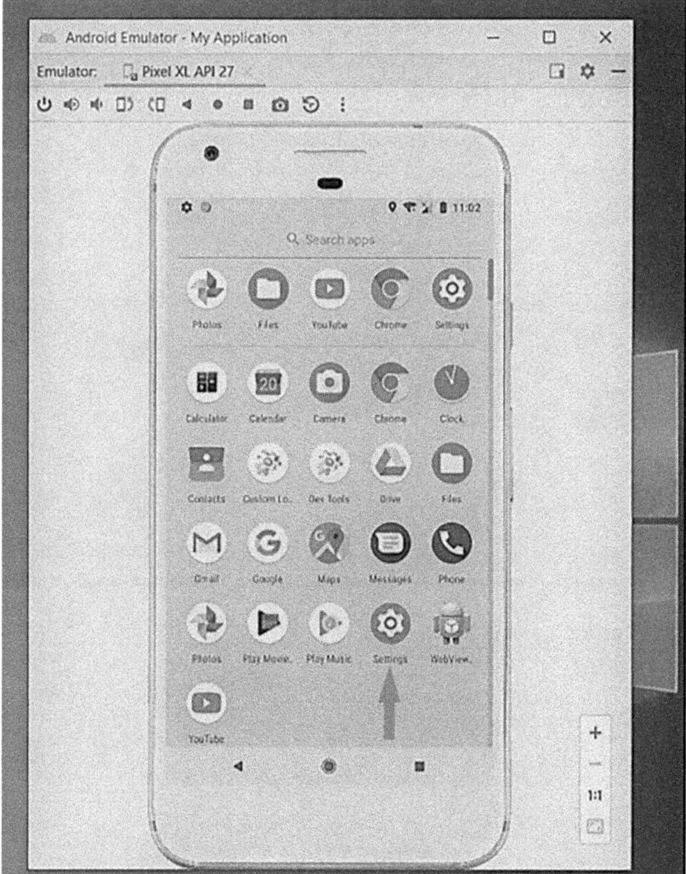

Figure 5.29 Opening "Settings" in the Android.

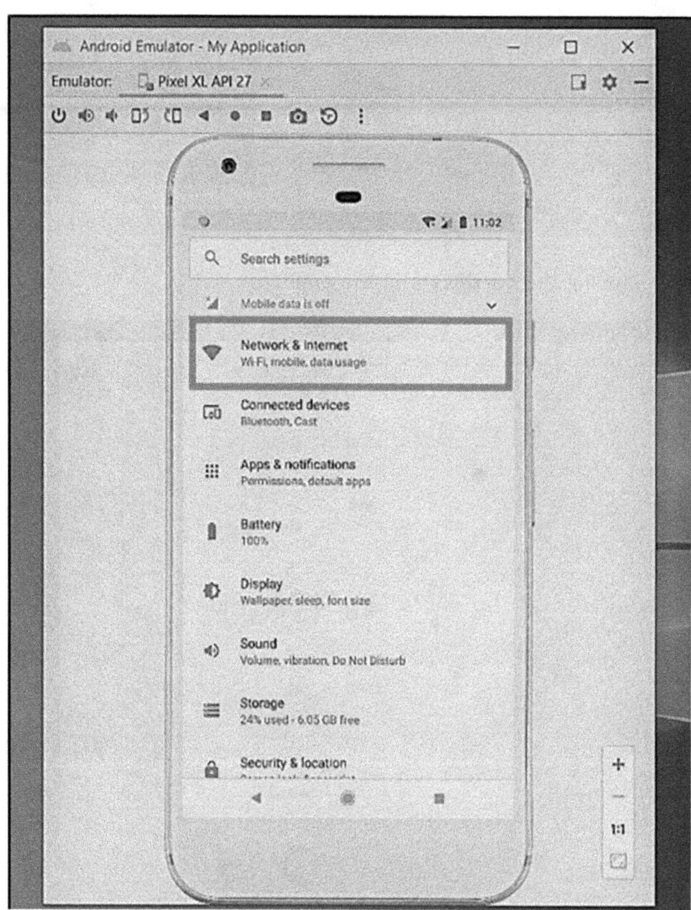

Figure 5.30 Setting up the certificates.

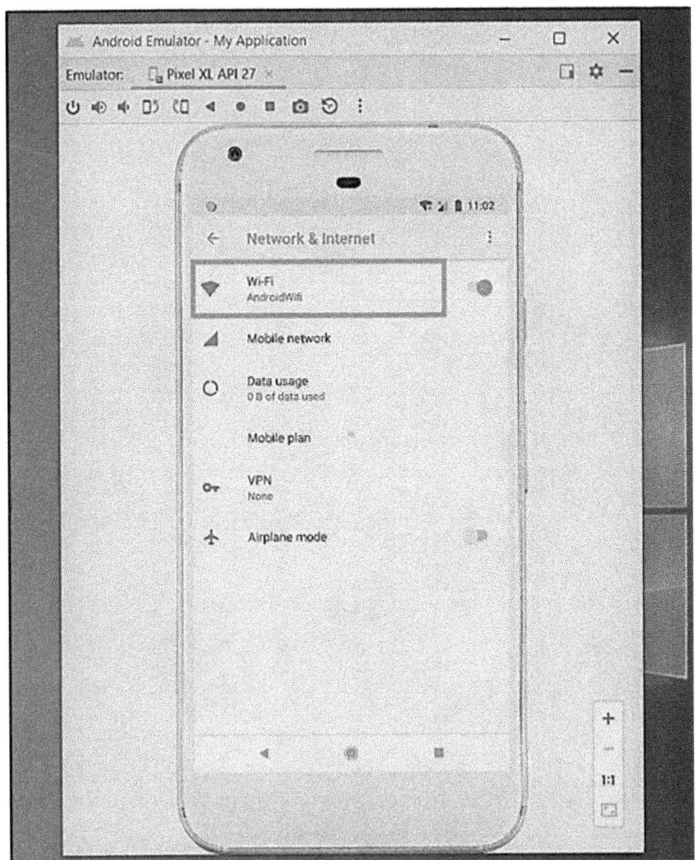

Figure 5.31 Importing the certificate in the emulator.

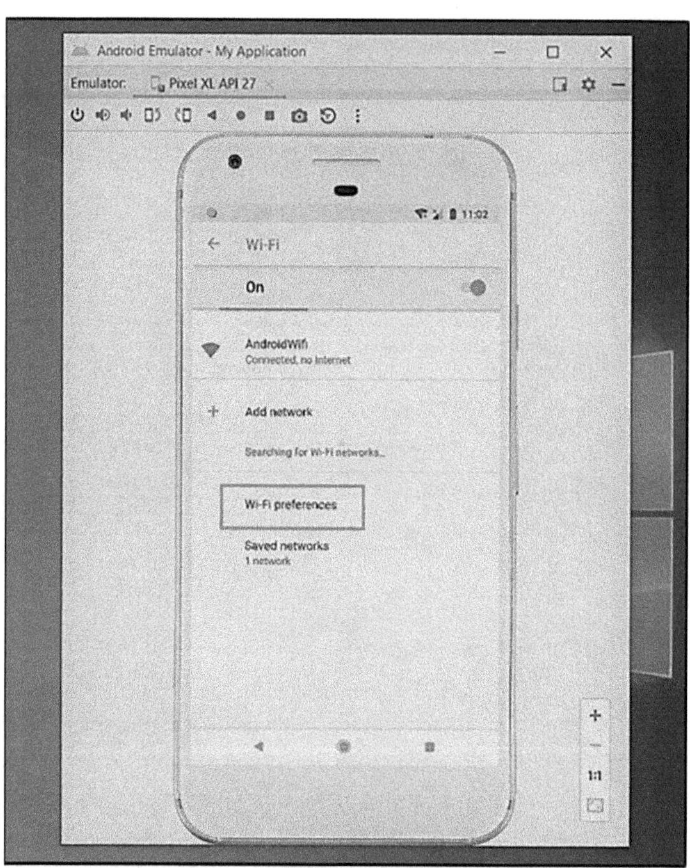

Figure 5.32 Installing the certificate in the emulator.

Step #13: Click on advanced options and select Install certificates.
Step #14: Locate the certificate we dragged and click on it.
Step #15: Give a certain name and click OK.
Step #16: Now, the phone might show some security- and privacy-related prompts. Go as per the methods demonstrated.
Now, depending upon what Android you are using, you might get this confirmation prompt regarding installing the certificate. Click ACCEPT to install the certificate.
Step #17: Now, after the certificate is installed, we have to configure the proxy settings in the Android emulator. So, for that go to Android Studio. Navigate to File > Settings > Appearance > System Settings > HTTP Proxy.
Step #18: Now, enable Manual Proxy Configuration, and under it, enable HTTP and set Hostname as 127.0.0.1 and Port as 8085 (as set in Burp Suite).

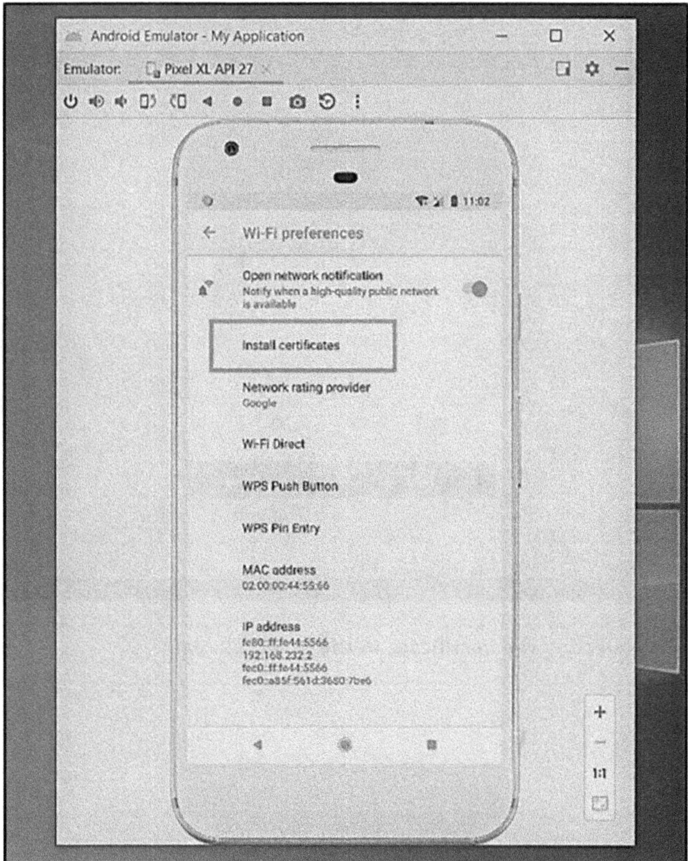

Figure 5.33 Installing the certificate in the emulator.

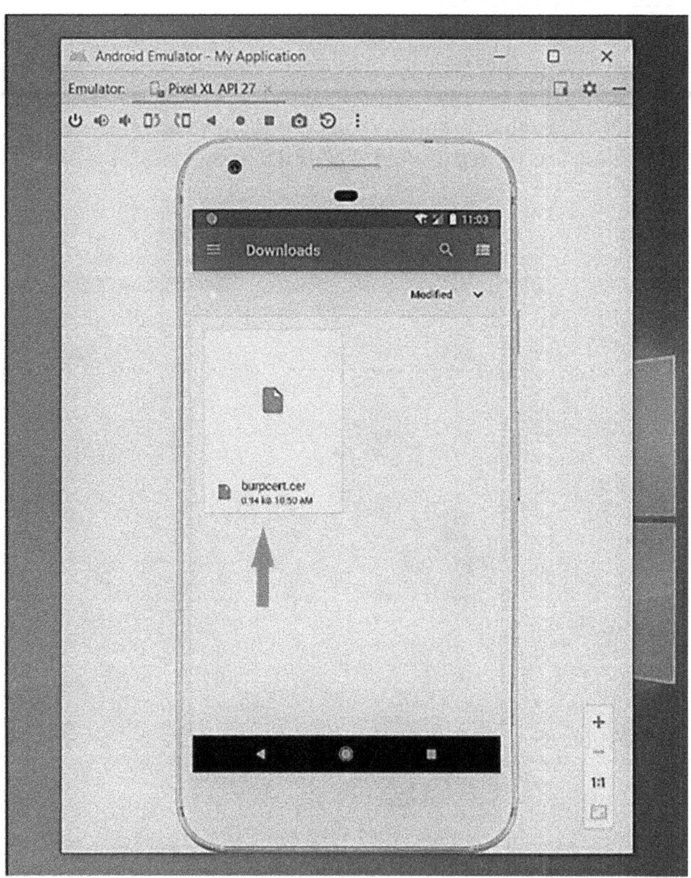

Figure 5.34 Selecting the certificate in the saved file path.

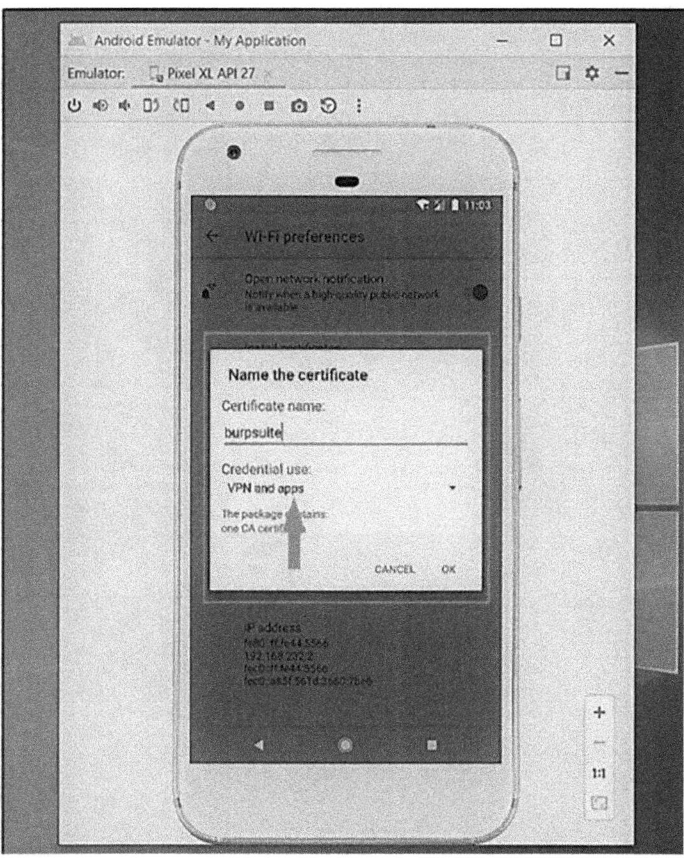

Figure 5.35 Installing the certificate in the emulator.

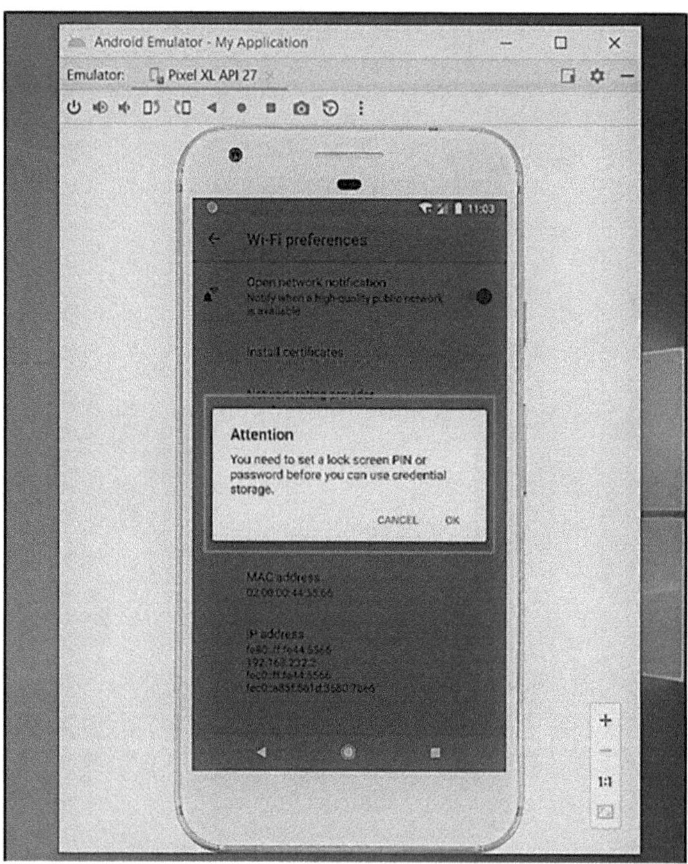

Figure 5.36 Click OK.

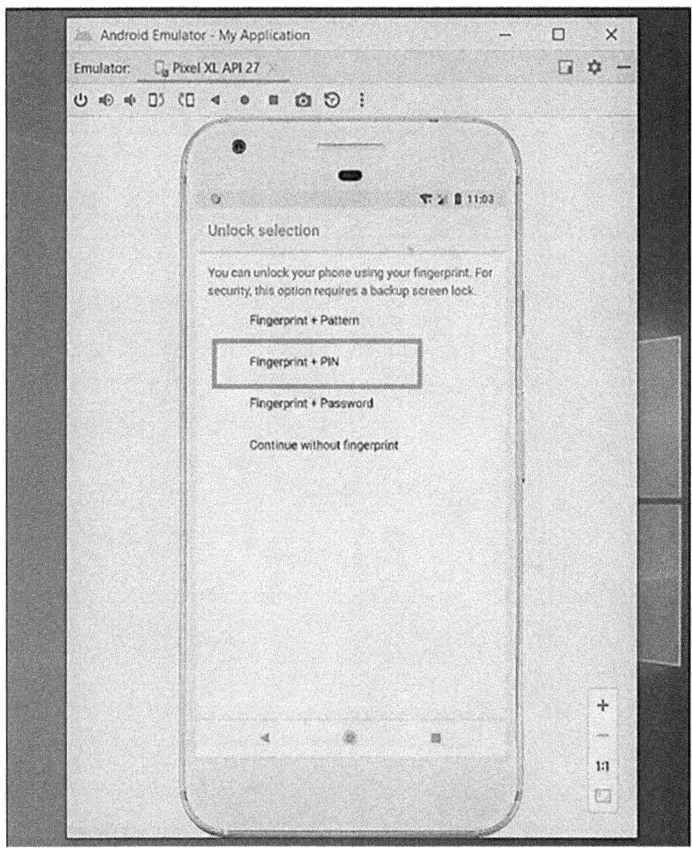

Figure 5.37 Click on any option, but PIN would be easier.

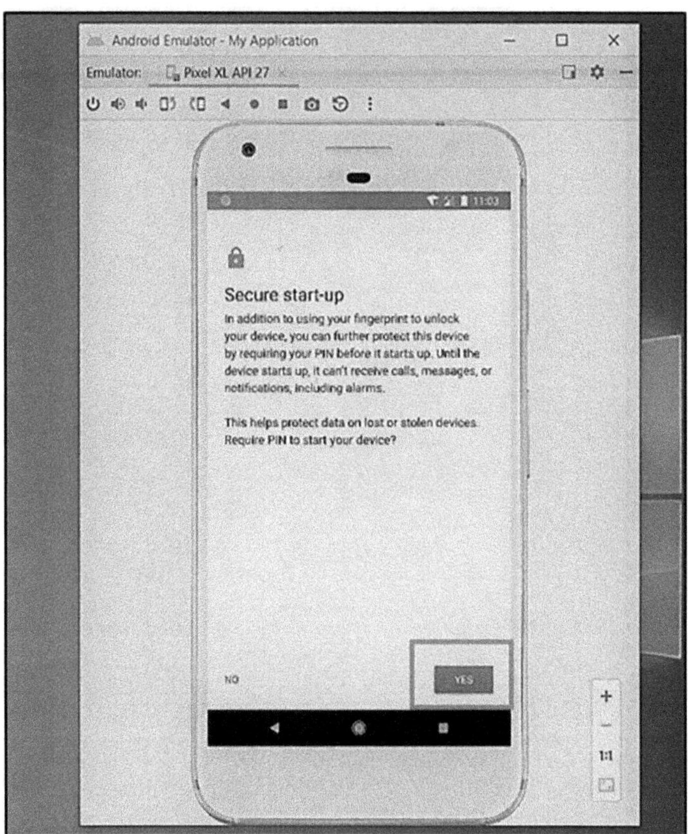

Figure 5.38 Click YES to start the setup.

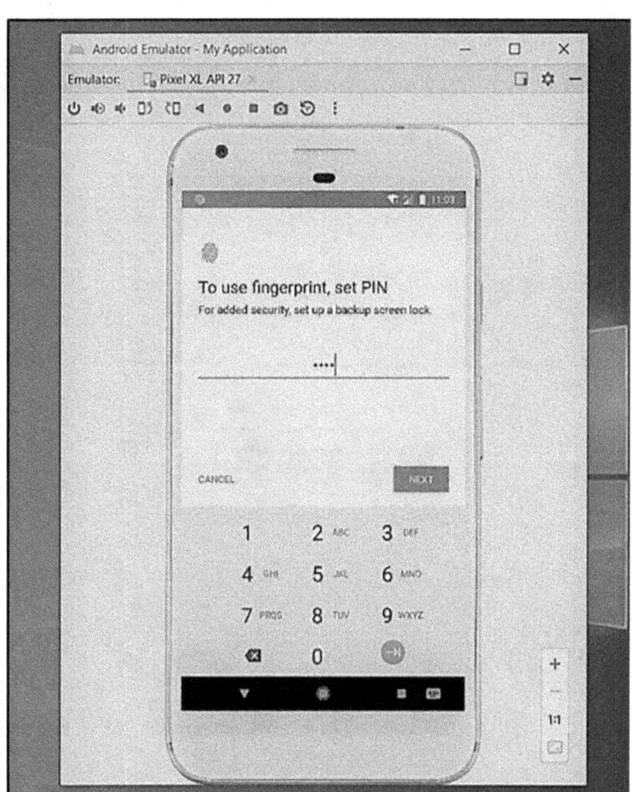

Figure 5.39 Give any PIN as it doesn't matter much.

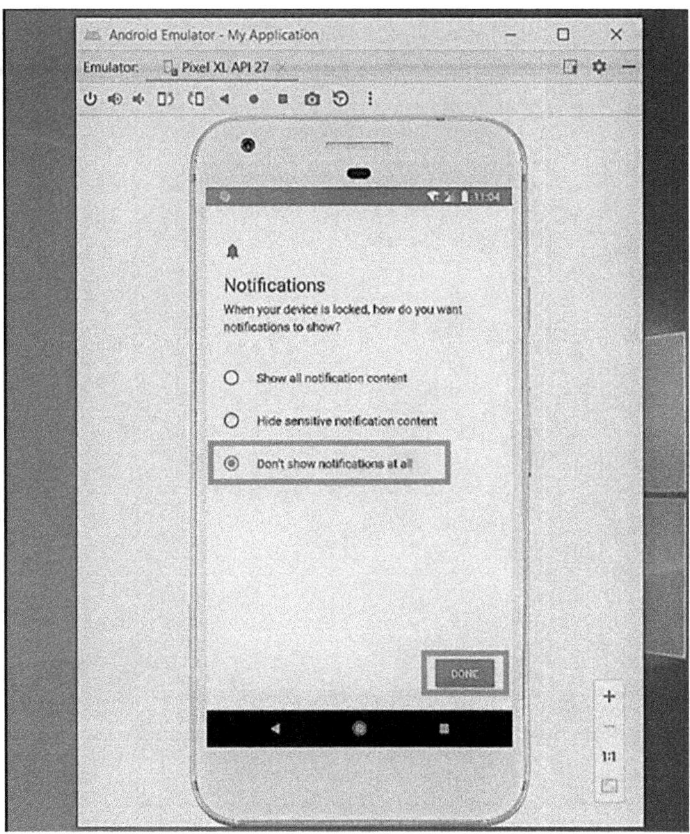

Figure 5.40 Click on the third option as these are for security purposes.

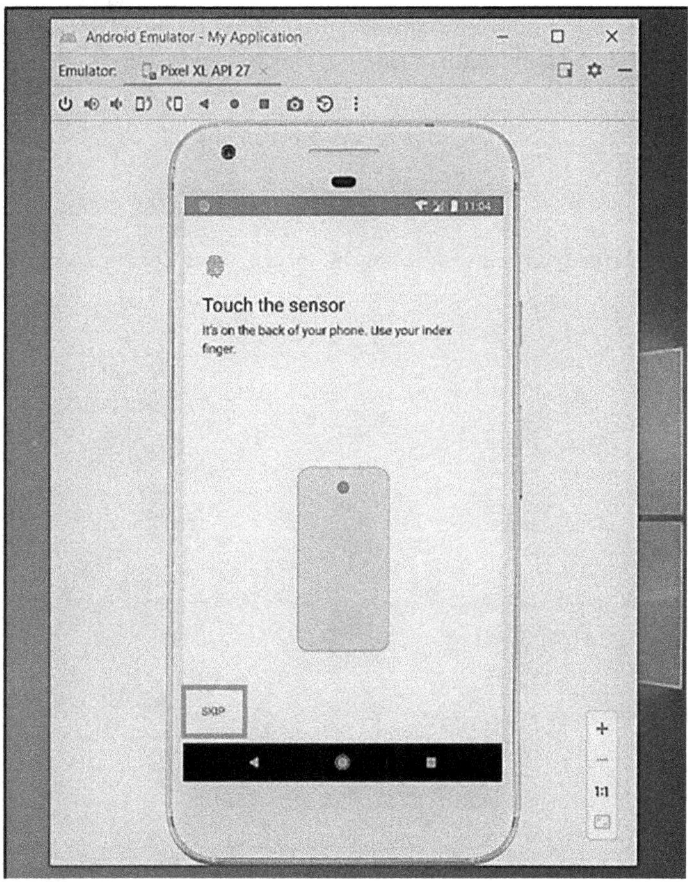

Figure 5.41 Click SKIP as it is an emulator.

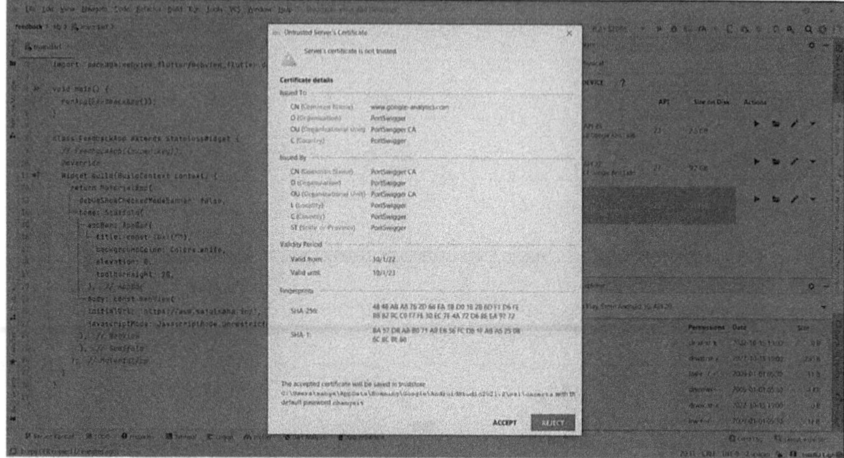

Figure 5.42 Security prompt regarding an unrecognized certificate installation.

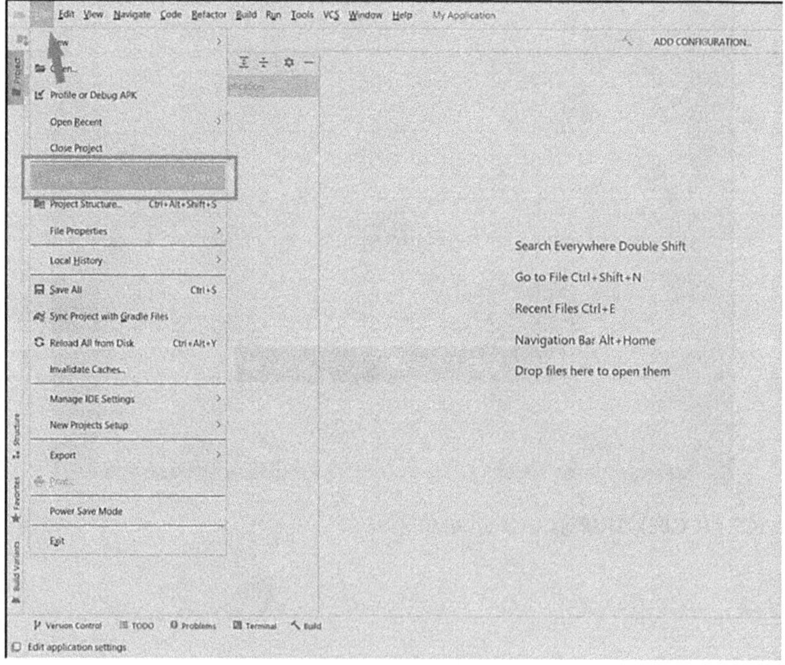

Figure 5.43 Setting up the proxy such that traffic goes through Burp Suite.

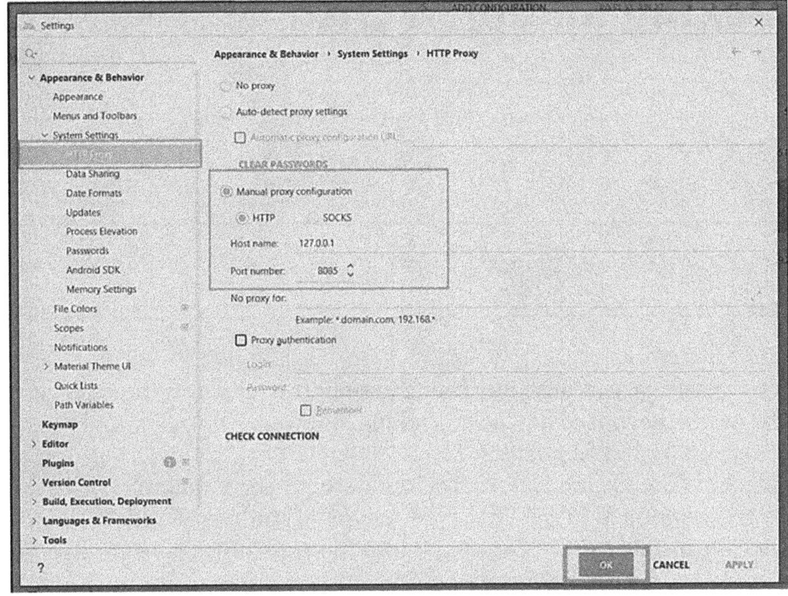

Figure 5.44 Configuring the proxy settings.

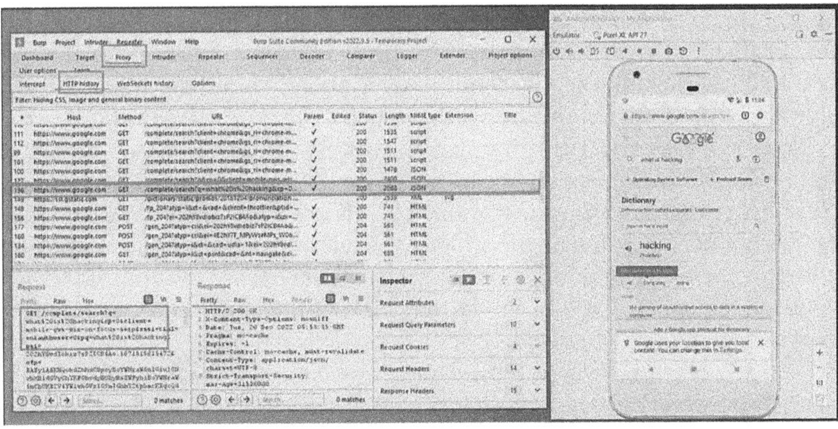

Figure 5.45 The traffic is passing via Burp Suite.

Step #19: Now, the golden time. To verify whether it is working, go to Burp Suite. Under the Proxy tab, go to the HTTP History tab opened. Go to the emulator and the browser. Search for anything and the traffic will get reflected in Burp's HTTP History Tab.

Great so our traffic is visible. Now, all setup has been done. Let's get into the keyboard for some cool hacking stuff.

Chapter 6

Hacking an app – Diva

Since everything is ready and running smoothly, it's now time to get our hands on the keyboard and start hacking. So, we need to first and foremost fire up the VM and Emulator.

Step #1: Connecting our Kali Machine with the emulator. The instructions are mentioned in the "Passive Analysis of Android Traffic" section.

Step #2: Installing the "Diva" apk into the emulator. There are two ways, either drag and drop or install via CLI. We are going to do it via the CLI. So, for that,

Type: adb -H <hostMachineIP> -P 5037 install Desktop/diva.apk

Now, if we open the app in our emulator, we get to see some challenges. Let's solve those one by one. At first, we got "Insecure Logging." Clicking on that challenge we get to see that we've been asked for the credit card number. And on clicking the Checkout button, some action will happen. Let's give a random number and see what happens.

Now, if we clearly read the objective of the challenge stated above, it states that Insecure logging occurs when developers intentionally or unintentionally log sensitive information. This means that the credit card number we will be supplying might get logged somewhere. So, we will be starting a Logcat server using adb that will give us all the logs of the android.

Logcat is basically a CLI utility that logs all the system messages whenever certain action is invoked.

Step #3: Starting Logcat server and providing the credit card number. Let's check what happens.

Now, there are various approaches to start and get logs of Android. We are going to discuss how to do it with the Kali Machine itself.

For that type: **adb -H <hostMachineIP> -P 5037 logcat**

This would start the logcat server, but the output of it may not be satisfying as these are logs and your entire terminal will be filled with it. This would solve the challenge though, but we are intended toward the efficient path.

DOI: 10.1201/9781003450474-6

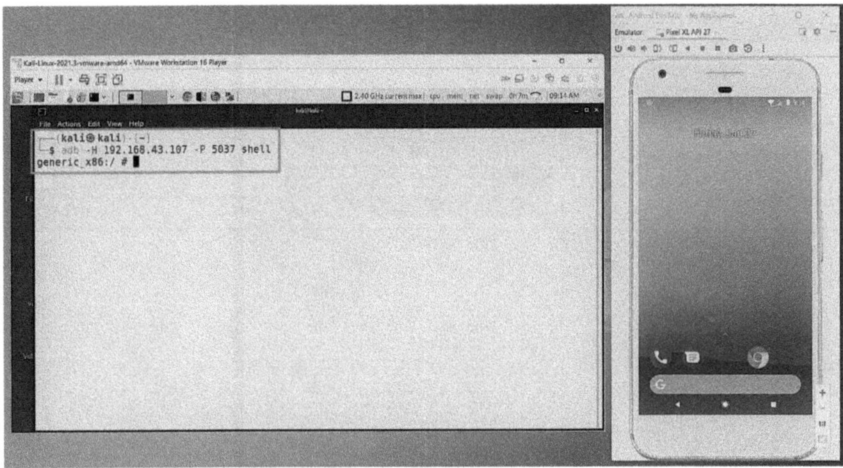

Figure 6.1 Connecting the Kali VM with the emulator.

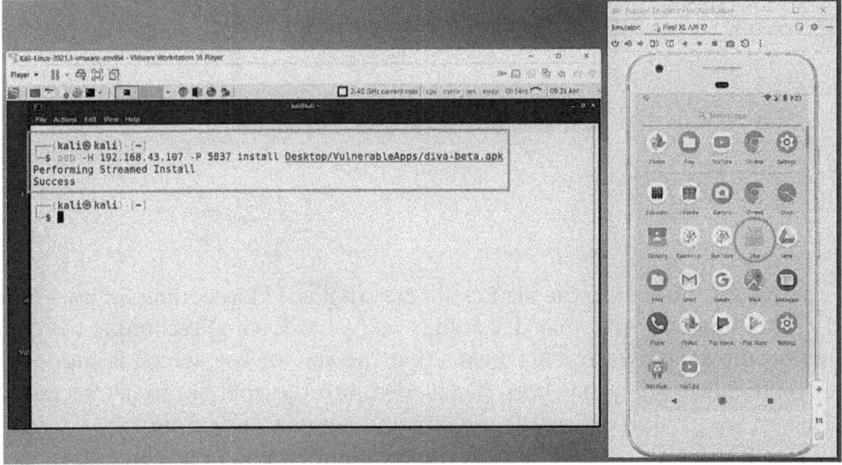

Figure 6.2 Installing the apk into the emulator.

Another great way would be to filter the cred we are looking for.

Type: **adb -H <hostMachineIP> -P 5037 logcat | grep <creditCardNumber>**

This would only show us the line where our credit card number is being logged

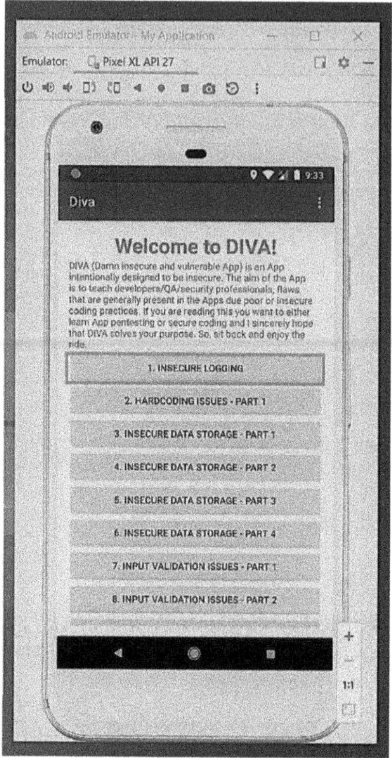

Figure 6.3 List of challenges.

Now, let's jump into the next challenge, that is, "Hardcoding Issues – Part 1." The objective says that developers use a hardcoding technique to fetch important data at ease. This means that the vendor key, which is supposed to be handled safely, has been hardcoded into the app. So, to get access to the key, we have to decompile the app and get its source code.

Step #1: Go to the VM and in the terminal type jadx-gui and go as per the procedure.

Step #2: The Jadx tool has been started. Click Open File, locate the target apk, and click Open.

Step #3: The app source code will be decompiled. Click any of the source files to see the code.

Step #4: Now, for the purpose of solving the challenge, head over to *Source Code > jakhar.aseem.diva > HardcodeActivity*. Click on it and you will be getting the source code.

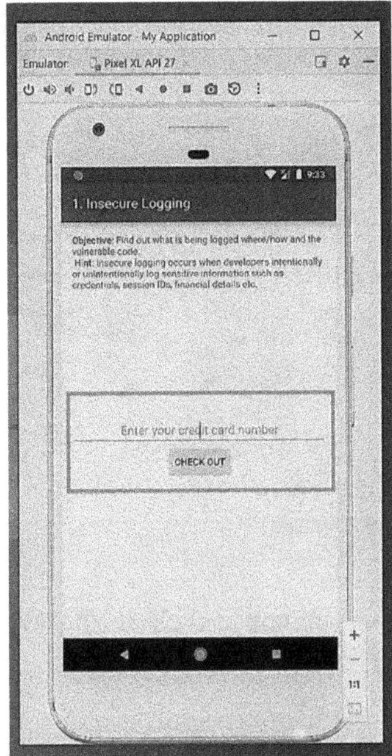

Figure 6.4 Challenge screen.

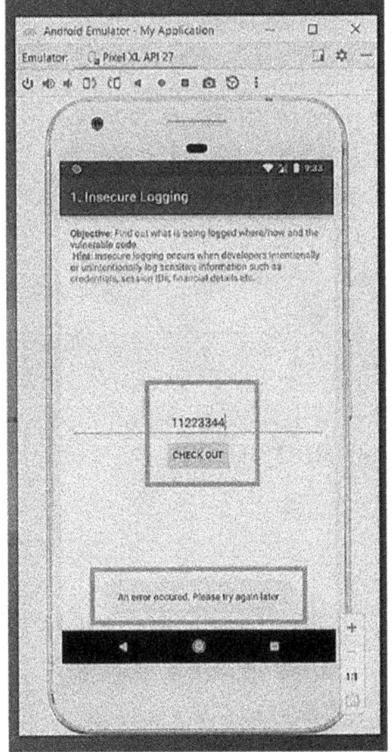

Figure 6.5 This is what happens.

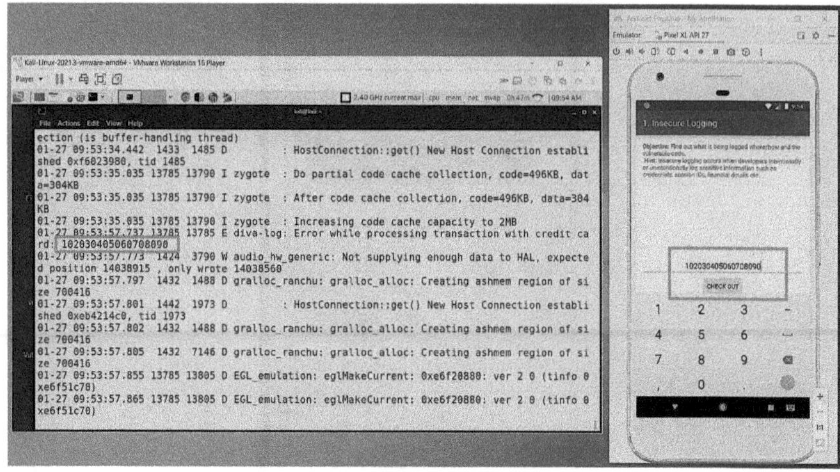

Figure 6.6 The logcat server running in the Kali VM reveals the logged credentials in a plain-text manner.

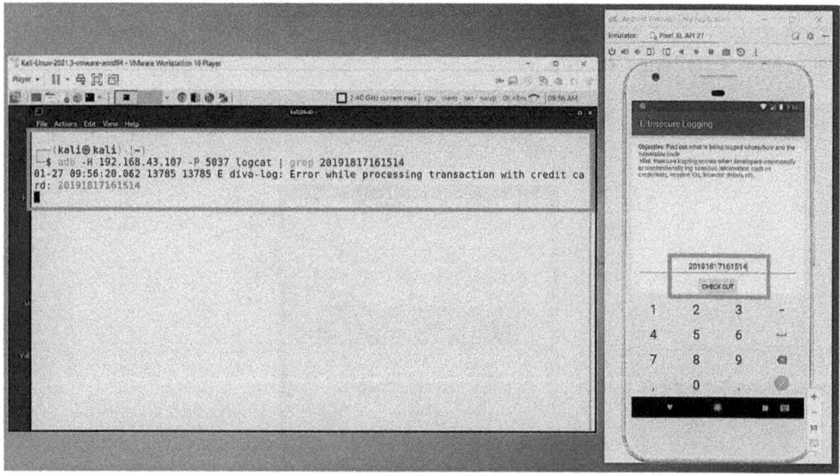

Figure 6.7 Filtering the logcat output.

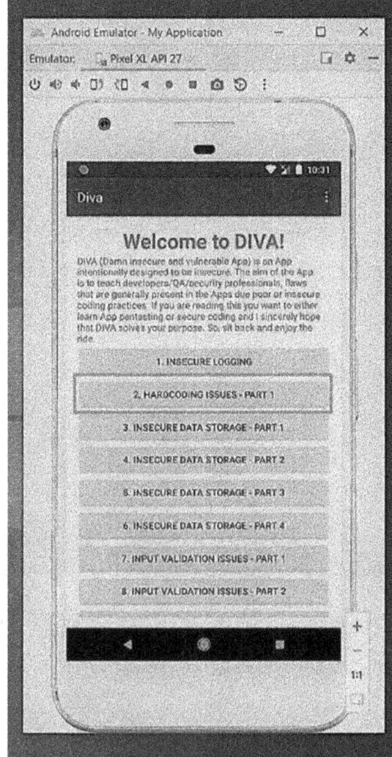

Figure 6.8 List of challenges.

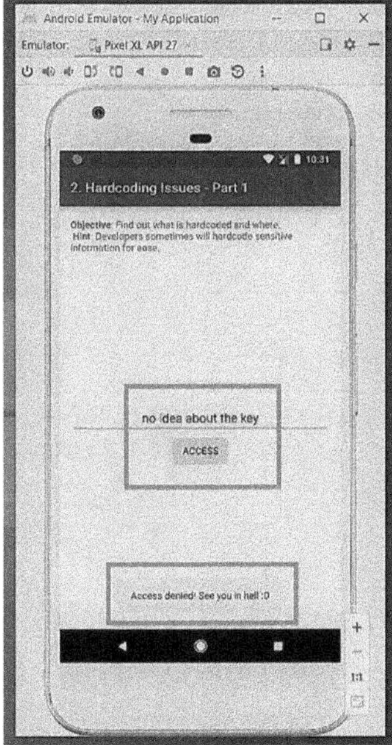

Figure 6.9 The challenge prompt.

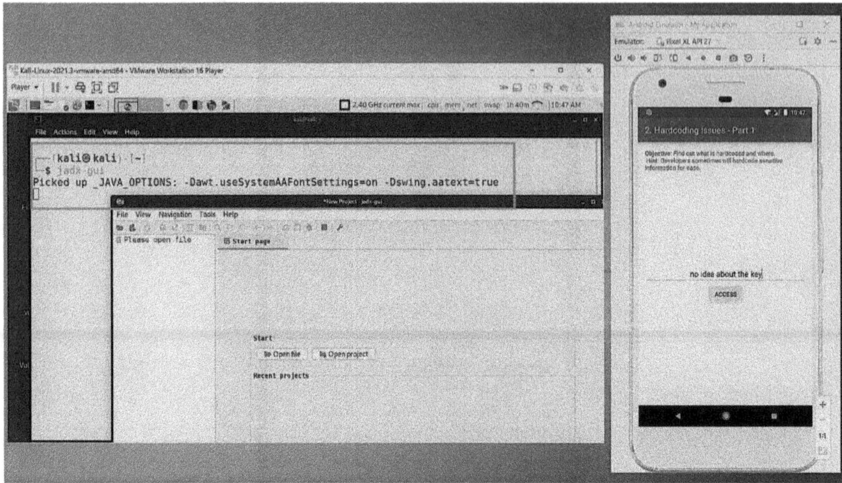

Figure 6.10 Starting Jadx-GUI in the VM for decompilation.

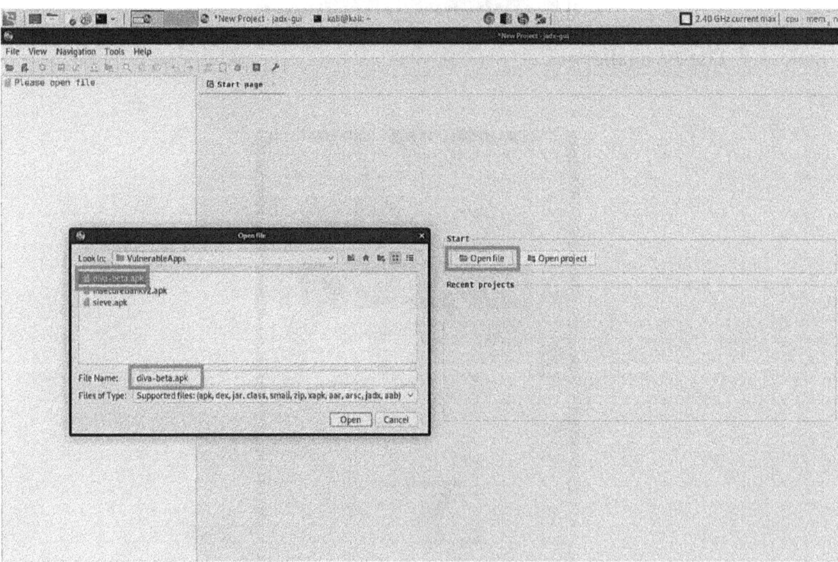

Figure 6.11 The Jadx-GUI window.

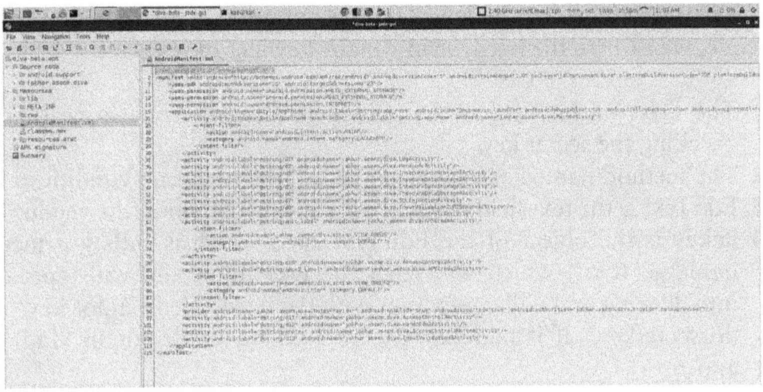

Figure 6.12 The decompiled application.

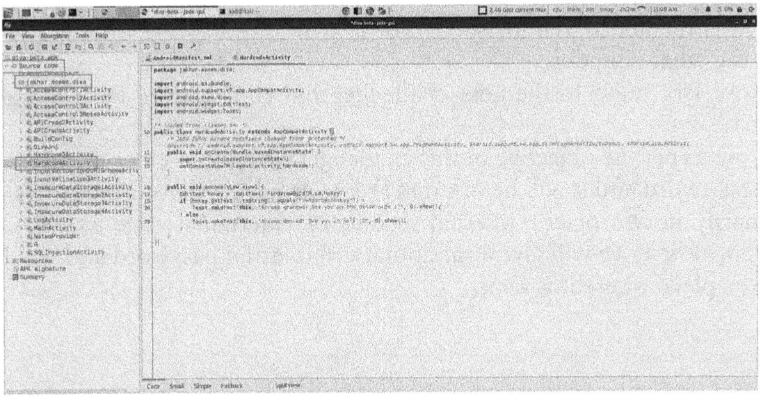

Figure 6.13 The challenge activity code.

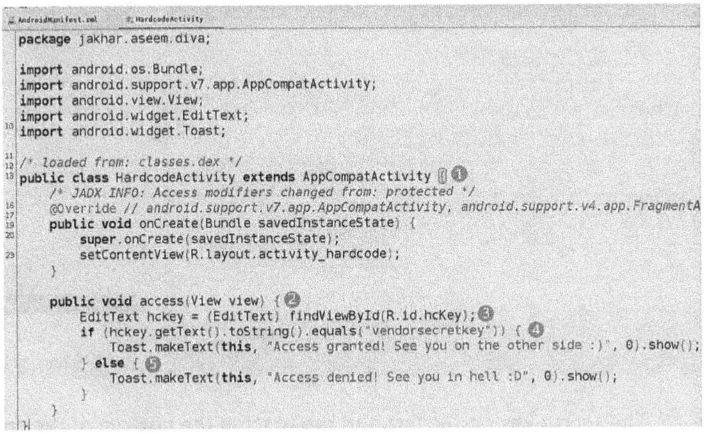

```
AndroidManifest.xml    HardcodeActivity
package jakhar.aseem.diva;

import android.os.Bundle;
import android.support.v7.app.AppCompatActivity;
import android.view.View;
import android.widget.EditText;
import android.widget.Toast;

/* loaded from: classes.dex */
public class HardcodeActivity extends AppCompatActivity {
    /* JADX INFO: Access modifiers changed from: protected */
    @Override // android.support.v7.app.AppCompatActivity, android.support.v4.app.FragmentA
    public void onCreate(Bundle savedInstanceState) {
        super.onCreate(savedInstanceState);
        setContentView(R.layout.activity_hardcode);
    }

    public void access(View view) {
        EditText hckey = (EditText) findViewById(R.id.hcKey);
        if (hckey.getText().toString().equals("vendorsecretkey")) {
            Toast.makeText(this, "Access granted! See you on the other side :)", 0).show();
        } else {
            Toast.makeText(this, "Access denied! See you in hell :D", 0).show();
        }
    }
}
```

Figure 6.14 List of challenges.

This is the code of the challenge activity. Now, if we properly go through the code, we can see the hardcoded vendor key. Let me explain.

1 The class within which the actual verification code is written for checking the secret key.
2 The method named access which has all the necessary conditions
3 EditText is the text field where we are giving the input (the vendor key)
4 hckey is the object of the EditText class which is calling a method name getText() which will fetch the user input and will typecast it into a String and will compare it with the hardcoded vendor key "vendorsecretkey." If it matches, then a toast (a small popup kind) will appear.
5 If the key is not equaling to the hardcoded key, then another toast will be shown, stating Access Denied.

Therefore, inside the conditional statement, the string that is being hardcoded is the actual vendor secret key that we are looking for. Let's give it and see what happens.

Now, let's jump to the next challenge, that is, "Insecure Data Storage – Part 1."

Now as per the objective of this challenge, we have to enter the username and password and it will be saved in the Android. Now, Android stores data primarily in two places: internal storage or external storage like SDCard, etc. So, at first, we will give a random username and password and will hunt for the place where it is kept.

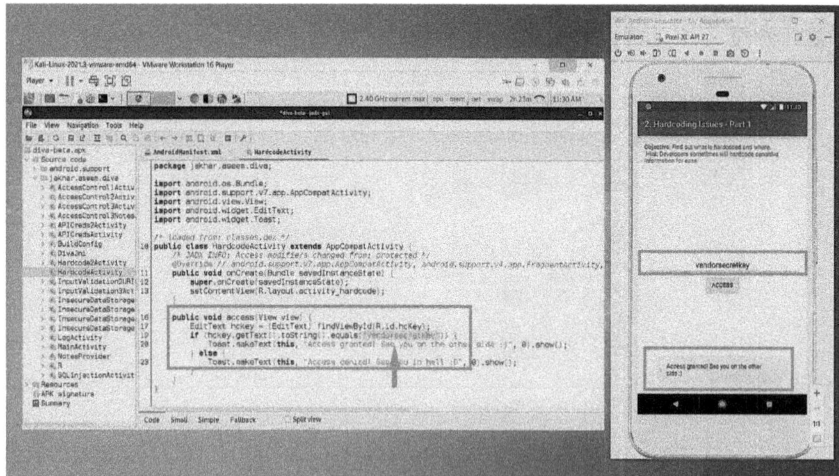

Figure 6.15 The secret key on the left, and the toast in the bottom of the emulator.

Figure 6.16 List of challenges.

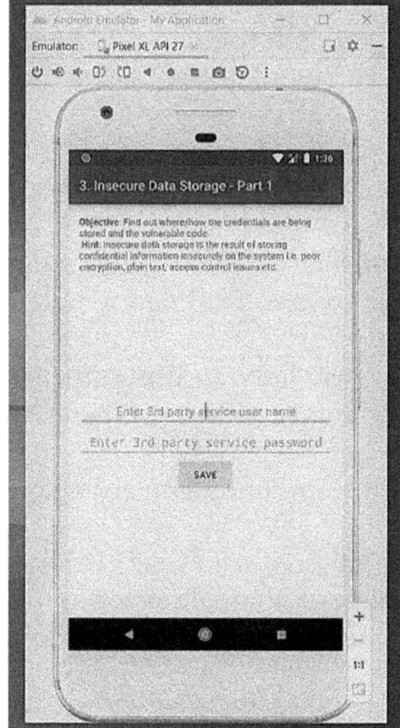

Figure 6.17 Main activity of the challenge.

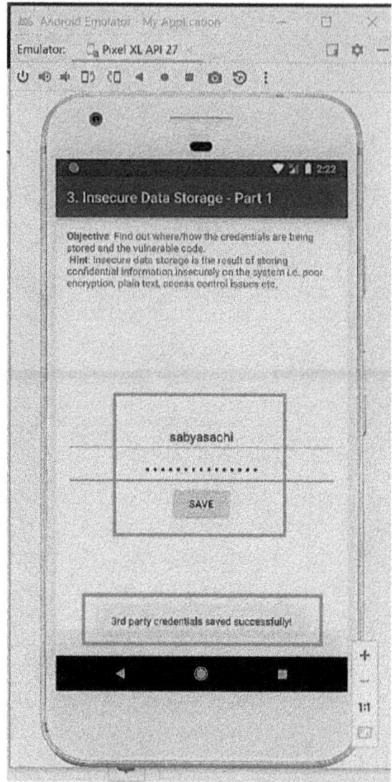

Figure 6.18 Inserting random username and password.

Step #1: Supply some random username and password to see the action happening. In this case, the username is “sabyasachi” and the password is “sabyasachi_pass.”

The toast says third-party credentials saved successfully. Now as discussed, Android can store data in either internal or external storage. External storage could be SDCard, and Internal storage could be some databases, files, libraries, caches, or Shared Preferences. A SharedPreferences object provides straightforward ways to read and write key-value pairs from and to files that contain them.

Step #2: In the VM, open Jadx and look the source code of the challenge activity. It is inside Source Code > jakhar.aseem.diva > InsecureData Storage1Activity

Alright, the code might seem intimidating. Let me explain.

1) The method where the credentials are getting stored
2) We can see and object of SharedPreferences names spref has been created. This could be the hint that our data is being stored in the Shared Preferences.

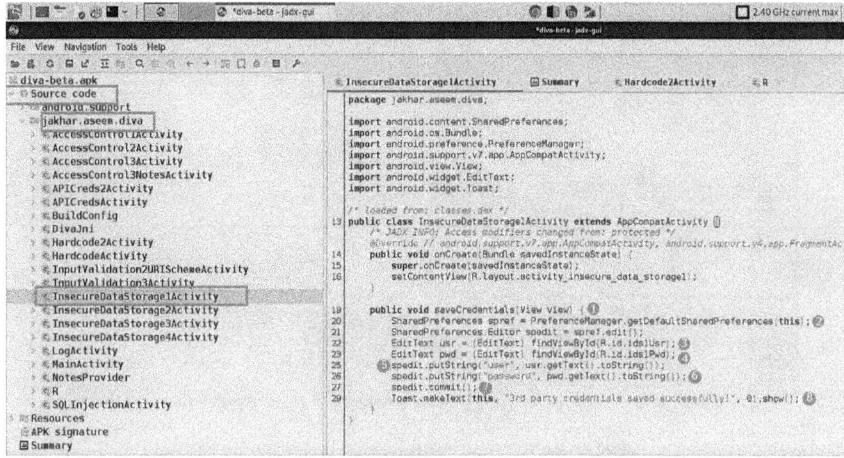

Figure 6.19 The source code of the challenge activity.

3) Text field named usr is created which will take our username as user input
4) Text field named pwd is created which will take our password as user input
5) spedit.putString basically taking the user input from the usr text field and converting it into a string format and storing it inside shared preferences
6) The same is happening over here as well.
7) Commit is saving the sharedpreferences.
8) Toast is given for us to understand that the creds have been successfully saved.

Note: If you are having difficulty understanding the code and what everything means, I would highly recommend learning Java and getting a basic Android Development experience as it would really help in building the foundation.

Step #3: In the VM, access the emulator

Note the # after emulator name, it means root level access.

Step #4: Go inside data/data/<package_name>/shared_pref/<filename>.xml

The shared preferences could be found inside /data/data/jakhar.aseem.diva/shared_pref. Inside, there will be a file named <package_name>_preferences.xml. If we cat the content of the xml file, we can see the credentials present inside it. Hence, challenge solved.

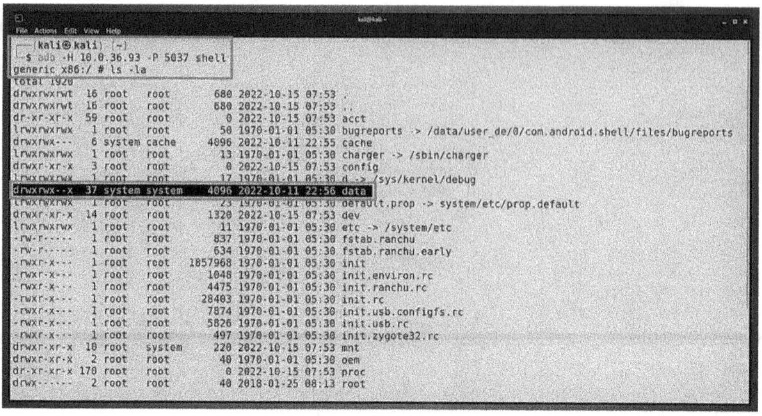

```
┌──(kali㉿kali)-[~]
└─$ adb -H 10.0.36.93 -P 5037 shell
generic_x86:/ # ls -la
total 1920
drwxrwxrwt  16 root   root        680 2022-10-15 07:53 .
drwxrwxrwt  16 root   root        680 2022-10-15 07:53 ..
dr-xr-xr-x  59 root   root          0 2022-10-15 07:53 acct
lrwxrwxrwx   1 root   root         50 1970-01-01 05:30 bugreports -> /data/user_de/0/com.android.shell/files/bugreports
drwxrwx---   6 system cache      4096 2022-10-11 22:55 cache
lrwxrwxrwx   1 root   root         13 1970-01-01 05:30 charger -> /sbin/charger
drwxr-xr-x   3 root   root          0 2022-10-15 07:53 config
lrwxrwxrwx   1 root   root         17 1970-01-01 05:30 d -> /sys/kernel/debug
drwxrwx--x  37 system system     4096 2022-10-11 22:56 data
lrwxrwxrwx   1 root   root         23 1970-01-01 05:30 default.prop -> system/etc/prop.default
drwxr-xr-x  14 root   root       1320 2022-10-15 07:53 dev
lrwxrwxrwx   1 root   root         11 1970-01-01 05:30 etc -> /system/etc
-rw-r-----   1 root   root        837 1970-01-01 05:30 fstab.ranchu
-rw-r-----   1 root   root        634 1970-01-01 05:30 fstab.ranchu.early
-rwxr-x---   1 root   root    1857968 1970-01-01 05:30 init
-rwxr-x---   1 root   root       1048 1970-01-01 05:30 init.environ.rc
-rwxr-x---   1 root   root       4475 1970-01-01 05:30 init.ranchu.rc
-rwxr-x---   1 root   root      28403 1970-01-01 05:30 init.rc
-rwxr-x---   1 root   root       7874 1970-01-01 05:30 init.usb.configfs.rc
-rwxr-x---   1 root   root       5826 1970-01-01 05:30 init.usb.rc
-rwxr-x---   1 root   root        497 1970-01-01 05:30 init.zygote32.rc
drwxr-xr-x  10 root   system      220 2022-10-15 07:53 mnt
drwxr-xr-x   2 root   root         40 1970-01-01 05:30 oem
dr-xr-xr-x 170 root   root          0 2022-10-15 07:53 proc
drwx------   2 root   root         40 2018-01-25 08:13 root
```

Figure 6.20 The source code of the challenge activity.

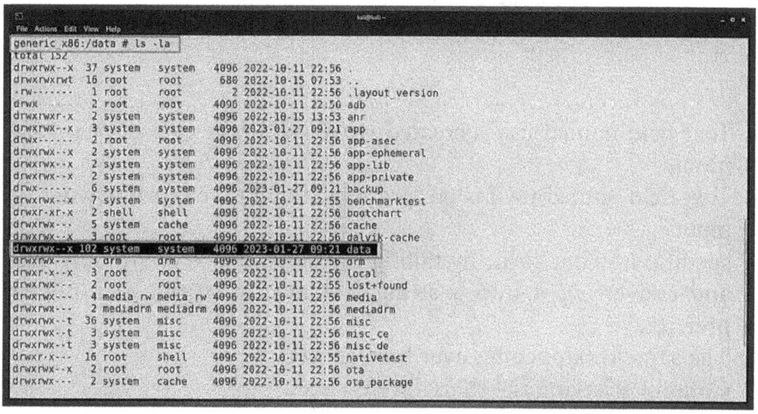

```
generic_x86:/data # ls -la
total 152
drwxrwx--x  37 system   system   4096 2022-10-11 22:56 .
drwxrwxrwt  16 root     root      680 2022-10-15 07:53 ..
-rw-------   1 root     root        2 2022-10-11 22:56 .layout_version
drwx------   2 root     root     4096 2022-10-11 22:50 adb
drwxrwxr-x   2 system   system   4096 2022-10-15 13:53 anr
drwxrwx--x   3 system   system   4096 2023-01-27 09:21 app
drwx------   2 root     root     4096 2022-10-11 22:56 app-asec
drwxrwx--x   2 system   system   4096 2022-10-11 22:56 app-ephemeral
drwxrwx--x   2 system   system   4096 2022-10-11 22:56 app-lib
drwxrwx--x   2 system   system   4096 2022-10-11 22:56 app-private
drwx------   6 system   system   4096 2023-01-27 09:21 backup
drwxrwx--x   7 system   system   4096 2022-10-11 22:55 benchmarktest
drwxr-xr-x   2 shell    shell    4096 2022-10-11 22:56 bootchart
drwxrwx---   5 system   cache    4096 2022-10-11 22:56 cache
drwxrwx--x   3 root     root     4096 2022-10-11 22:56 dalvik-cache
drwxrwx--x 102 system   system   4096 2023-01-27 09:21 data
drwxrwx---   3 drm      drm      4096 2022-10-11 22:56 drm
drwxr-x--x   3 root     root     4096 2022-10-11 22:56 local
drwxrwx---   2 root     root     4096 2022-10-11 22:55 lost+found
drwxrwx---   4 media_rw media_rw 4096 2022-10-11 22:56 media
drwxrwx---   2 mediadrm mediadrm 4096 2022-10-11 22:56 mediadrm
drwxrwx--t  36 system   misc     4096 2022-10-11 22:56 misc
drwxrwx--t   3 system   misc     4096 2022-10-11 22:56 misc_ce
drwxrwx--t   3 system   misc     4096 2022-10-11 22:56 misc_de
drwxr-x---  16 root     shell    4096 2022-10-11 22:55 nativetest
drwxrwx--x   2 root     root     4096 2022-10-11 22:56 ota
drwxrwx---   2 system   cache    4096 2022-10-11 22:56 ota_package
```

Figure 6.21 Inside the Android filesystem (inside /data/).

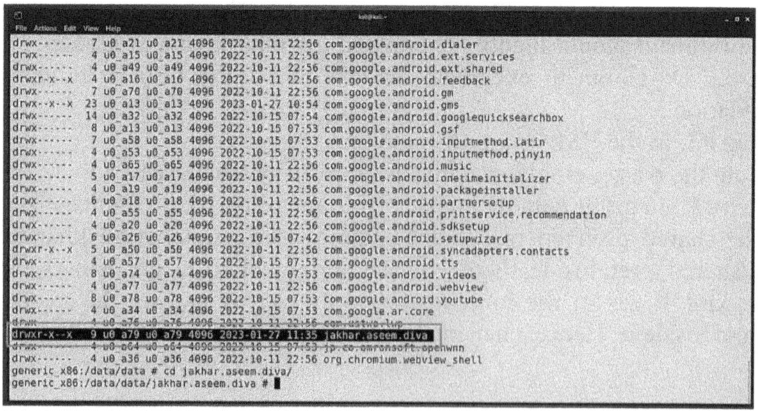

```
drwx------   7 u0_a21 u0_a21 4096 2022-10-11 22:56 com.google.android.dialer
drwx------   4 u0_a15 u0_a15 4096 2022-10-11 22:56 com.google.android.ext.services
drwx------   4 u0_a49 u0_a49 4096 2022-10-11 22:56 com.google.android.ext.shared
drwxr-x--x   4 u0_a16 u0_a16 4096 2022-10-11 22:56 com.google.android.feedback
drwx------   7 u0_a70 u0_a70 4096 2022-10-11 22:56 com.google.android.gm
drwx--x--x  23 u0_a13 u0_a13 4096 2023-01-27 10:54 com.google.android.gms
drwx------  14 u0_a32 u0_a32 4096 2022-10-15 07:54 com.google.android.googlequicksearchbox
drwx------   8 u0_a13 u0_a13 4096 2022-10-15 07:53 com.google.android.gsf
drwx------   7 u0_a58 u0_a58 4096 2022-10-15 07:53 com.google.android.inputmethod.latin
drwx------   4 u0_a53 u0_a53 4096 2022-10-15 07:53 com.google.android.inputmethod.pinyin
drwx------   4 u0_a65 u0_a65 4096 2022-10-11 22:56 com.google.android.music
drwx------   5 u0_a17 u0_a17 4096 2022-10-11 22:56 com.google.android.onetimeinitializer
drwx------   4 u0_a19 u0_a19 4096 2022-10-11 22:56 com.google.android.packageinstaller
drwx------   6 u0_a18 u0_a18 4096 2022-10-11 22:56 com.google.android.partnersetup
drwx------   4 u0_a55 u0_a55 4096 2022-10-11 22:56 com.google.android.printservice.recommendation
drwx------   4 u0_a20 u0_a20 4096 2022-10-11 22:56 com.google.android.sdksetup
drwx------   6 u0_a26 u0_a26 4096 2022-10-15 07:42 com.google.android.setupwizard
drwxr-x--x   5 u0_a50 u0_a50 4096 2022-10-11 22:56 com.google.android.syncadapters.contacts
drwx------   4 u0_a57 u0_a57 4096 2022-10-15 07:53 com.google.android.tts
drwx------   8 u0_a74 u0_a74 4096 2022-10-15 07:53 com.google.android.videos
drwx------   4 u0_a77 u0_a77 4096 2022-10-11 22:56 com.google.android.webview
drwx------   8 u0_a78 u0_a78 4096 2022-10-15 07:53 com.google.android.youtube
drwx------   4 u0_a34 u0_a34 4096 2022-10-15 07:53 com.google.ar.core
drwx------   4 u0_a76 u0_a76 4096 2022-10-11 22:56 com.ustwo.lwp
drwxr-x--x   9 u0_a79 u0_a79 4096 2023-01-27 11:35 jakhar.aseem.diva
drwx------   4 u0_a64 u0_a64 4096 2022-10-15 07:53 jp.co.omronsoft.openwnn
drwx------   4 u0_a36 u0_a36 4096 2022-10-11 22:56 org.chromium.webview_shell
generic_x86:/data/data # cd jakhar.aseem.diva/
generic_x86:/data/data/jakhar.aseem.diva #
```

Figure 6.22 Inside the Android filesystem (/data/data/<package_name>/).

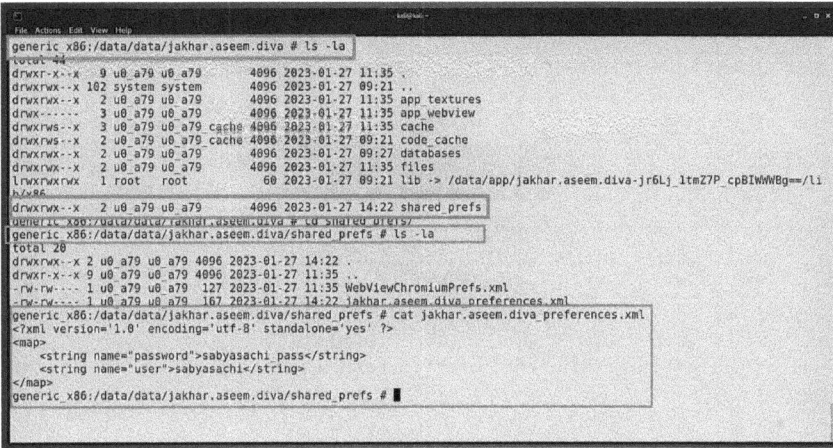

```
generic_x86:/data/data/jakhar.aseem.diva # ls -la
total 44
drwxr-x--x   9 u0_a79 u0_a79       4096 2023-01-27 11:35 .
drwxrwx--x 102 system system       4096 2023-01-27 09:21 ..
drwxrwx--x   2 u0_a79 u0_a79       4096 2023-01-27 11:35 app_textures
drwx------   3 u0_a79 u0_a79       4096 2023-01-27 11:35 app_webview
drwxrws--x   3 u0_a79 u0_a79_cache 4096 2023-01-27 11:35 cache
drwxrws--x   2 u0_a79 u0_a79_cache 4096 2023-01-27 09:21 code_cache
drwxrwx--x   2 u0_a79 u0_a79       4096 2023-01-27 09:27 databases
drwxrwx--x   2 u0_a79 u0_a79       4096 2023-01-27 11:35 files
lrwxrwxrwx   1 root   root           60 2023-01-27 09:21 lib -> /data/app/jakhar.aseem.diva-jr6Lj_ltmZ7P_cpBIWWWBg==/li
b/x86
drwxrwx--x   2 u0_a79 u0_a79       4096 2023-01-27 14:22 shared_prefs
generic_x86:/data/data/jakhar.aseem.diva # cd shared_prefs/
generic_x86:/data/data/jakhar.aseem.diva/shared_prefs # ls -la
total 20
drwxrwx--x 2 u0_a79 u0_a79 4096 2023-01-27 14:22 .
drwxr-x--x 9 u0_a79 u0_a79 4096 2023-01-27 11:35 ..
-rw-rw---- 1 u0_a79 u0_a79  127 2023-01-27 11:35 WebViewChromiumPrefs.xml
-rw-rw---- 1 u0_a79 u0_a79  167 2023-01-27 14:22 jakhar.aseem.diva_preferences.xml
generic_x86:/data/data/jakhar.aseem.diva/shared_prefs # cat jakhar.aseem.diva_preferences.xml
<?xml version='1.0' encoding='utf-8' standalone='yes' ?>
<map>
    <string name="password">sabyasachi_pass</string>
    <string name="user">sabyasachi</string>
</map>
generic_x86:/data/data/jakhar.aseem.diva/shared_prefs #
```

Figure 6.23 Inside the Android filesystem (/data/data/jakhar.aseem.diva/shared_prefs/).

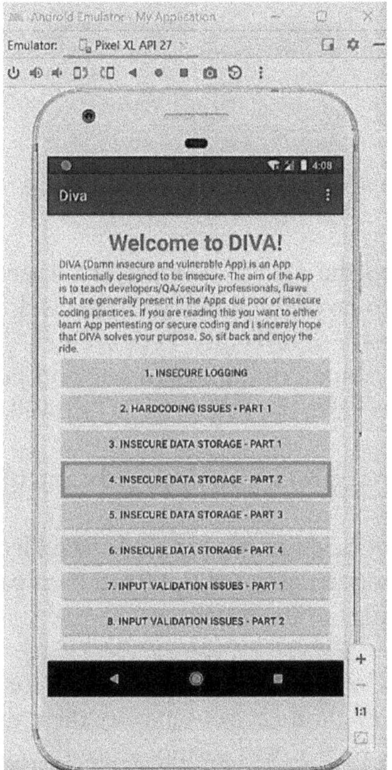

Figure 6.24 List of challenges.

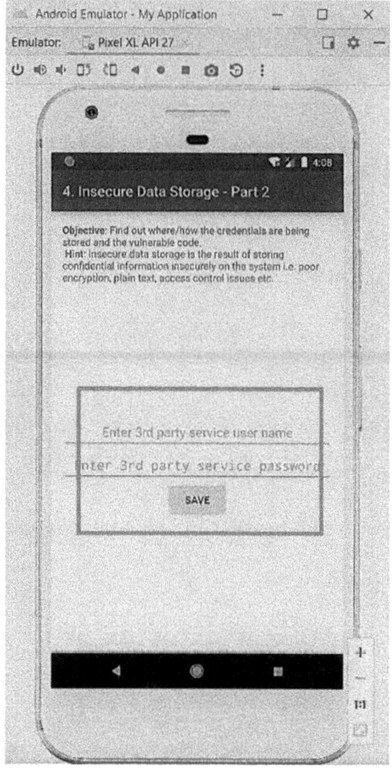

Figure 6.25 Challenge activity.

Now, let's jump into the next challenge. Try doing it yourself first. The next challenge seems to be of the same type, that is, Insecure Data Storage – Part 2 with some increase of difficulty. Let's see it.

Step #1: Give input some random username and password as of the previous challenge. In this case, the username is "sabya1" and the password is "sabya_pass1."

Step #2: Similarly, in the Kali VM, open Jadx and open the source code of the challenge activity.

Now, the code might be daunting for some. Well let me explain.

At first, the app is importing the SQLiteDatabase package so that the app could save the credentials in the database.

1 The method onCreate() creates a database for the app to store the credentials. The method checks whether there exists any table named myuser. If exists, then it will simply store the creds; else it will create a database and will start storing data into it. Inside the double quotes (" ") written is the SQL query to create database.
2 The method saveCredentials() saves the creds into the database. usr and pwd are the objects of the EditText class. EditText is the text field,

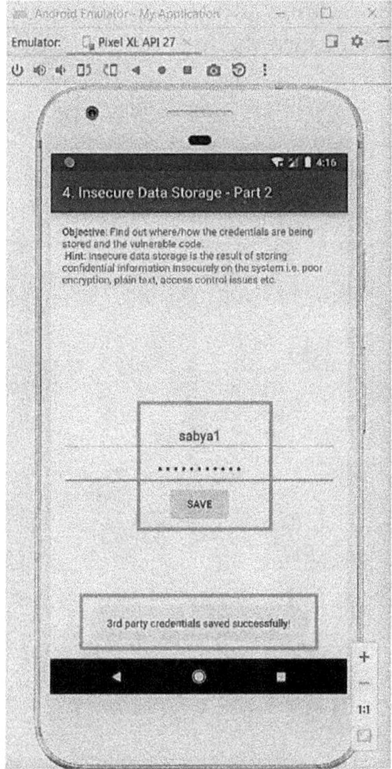

Figure 6.26 Giving random user input.

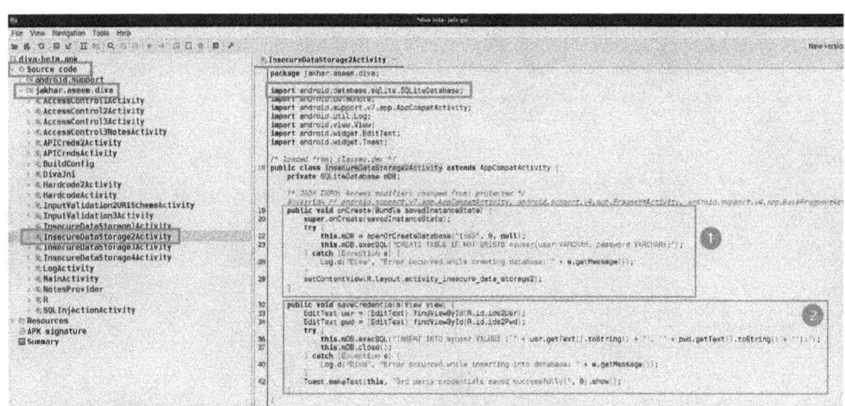

Figure 6.27 Source code of the challenge activity.

which takes the username and password as user inputs. Then taking those values, it puts it into the SQLite database using SQL Queries.

Therefore, we can understand that the creds are present in the databases. So, let's check where the database is located.

```
┌──(kali㉿kali)-[~]
└─$ adb -H 10.0.49.120 -P 5037 shell
generic_x86:/ # ls
acct       d              fstab.ranchu       init.rc                oem    storage            var
bugreports data           fstab.ranchu.early init.usb.configfs.rc   proc   sys                vendor
cache      default.prop   init               init.usb.rc            root   system
charger    dev            init.environ.rc    init.zygote32.rc       sbin   ueventd.ranchu.rc
config     etc            init.ranchu.rc     mnt                    sdcard ueventd.rc
generic_x86:/ # cd data/data/jakhar.aseem.diva/
generic_x86:/data/data/jakhar.aseem.diva # ls
app_textures app_webview cache code_cache databases files lib shared_prefs
generic_x86:/data/data/jakhar.aseem.diva # cd databases/
generic_x86:/data/data/jakhar.aseem.diva/databases # ls
divanotes.db divanotes.db-journal ids2 ids2-journal
generic_x86:/data/data/jakhar.aseem.diva/databases # ls -la
total 48
drwxrwx--x 2 u0_a79 u0_a79  4096 2023-01-27 15:56 .
drwxr-x--x 9 u0_a79 u0_a79  4096 2023-01-27 11:35 ..
-rw-rw---- 1 u0_a79 u0_a79 20480 2023-01-27 09:27 divanotes.db
-rw-rw---- 1 u0_a79 u0_a79     0 2023-01-27 09:27 divanotes.db-journal
-rw-rw---- 1 u0_a79 u0_a79 16384 2023-01-27 16:16 ids2
-rw-rw---- 1 u0_a79 u0_a79     0 2023-01-27 16:16 ids2-journal
generic_x86:/data/data/jakhar.aseem.diva/databases # cat ids2
Z0ZKstablemyusermyuserCREATE TABLE myuser(user VARCHAR, password VARCHAR)W--ctableandroid_metadataandroid_m
00#sabyalsabya pass1generic_x86:/data/data/jakhar.aseem.diva/databases #
```

Figure 6.28 Accessing the SQLite database.

```
generic_x86:/data/data/jakhar.aseem.diva/databases # ls
divanotes.db divanotes.db-journal ids2 ids2-journal
generic_x86:/data/data/jakhar.aseem.diva/databases # pwd
/data/data/jakhar.aseem.diva/databases
generic_x86:/data/data/jakhar.aseem.diva/databases # exit

┌──(kali㉿kali)-[~]
└─$ adb -H 10.0.49.120 -P 5037 pull /data/data/jakhar.aseem.diva/databases/ids2
/data/data/jakhar.aseem.diva/databases/ids2: 1 file pulled, 0 skipped. 5.1 MB/s (16384 bytes in 0.003s)

┌──(kali㉿kali)-[~]
└─$ ls
Desktop  Documents  Downloads  ids2  Music  Pictures  Public  Templates  Videos

┌──(kali㉿kali)-[~]
└─$ file ids2
ids2: SQLite 3.x database, last written using SQLite version 3019004

┌──(kali㉿kali)-[~]
└─$
```

Figure 6.29 Downloading the SQLite database.

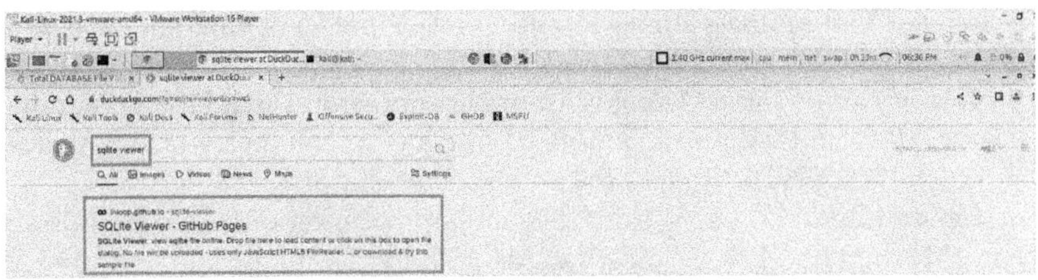

Figure 6.30 Accessing the SQLite database.

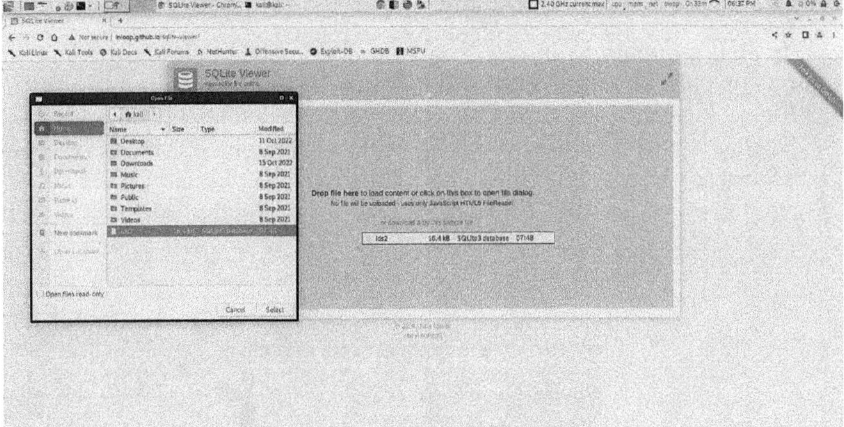

Figure 6.31 Accessing the SQLite database online.

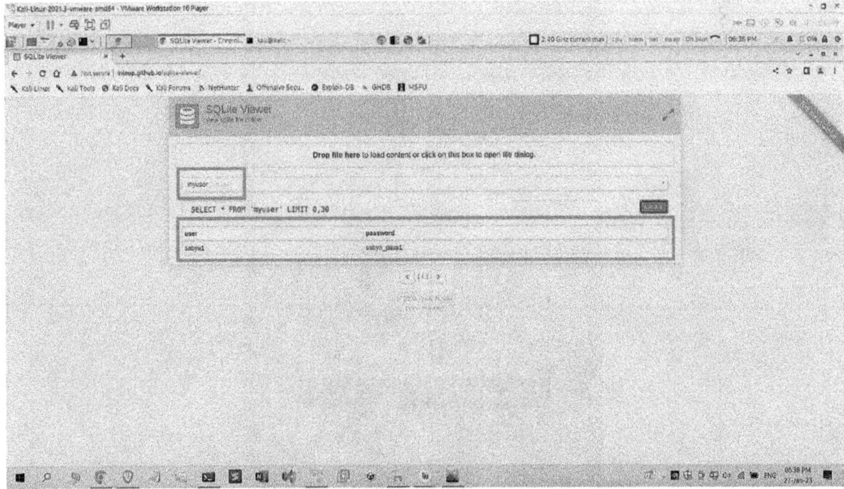

Figure 6.32 Reading the data from the SQLite database.

Step #3: The database folder is present inside the /data/data/jakhar.aseem.diva/databases. There's a file named ids2.

Now, it can be intimidating if an application stores database containing lots of tables and within each table having lots of columns which is typical in a real-life scenario. So, let's see an optimal solution.

Step #4 (Optional): Exit from the emulator connection and type: adb -H <hostMachineIP> -P 5037 pull /data/data/jakhar.aseem.diva/databases/ids2.

This command will basically pull the ids2 file, which is an SQLite database file in our Kali VM, and with the help of Online SQLite Viewer, we'll open the file.

Step #5 (Optional): Go to Browser, search for Sqlite Viewer, click on the highlighted link.

Step #6 (Optional): Drag and drop your ids2 SqliteDB file and select your table as myuser. The credentials will be listed.

Alright! So, this challenge has been completed. Let's jump into another one.

The next challenge deals with the same vulnerability, that is, Insecure Data Storage – Part 3, probably with a little twist. Let's see!

Figure 6.33 Challenge list.

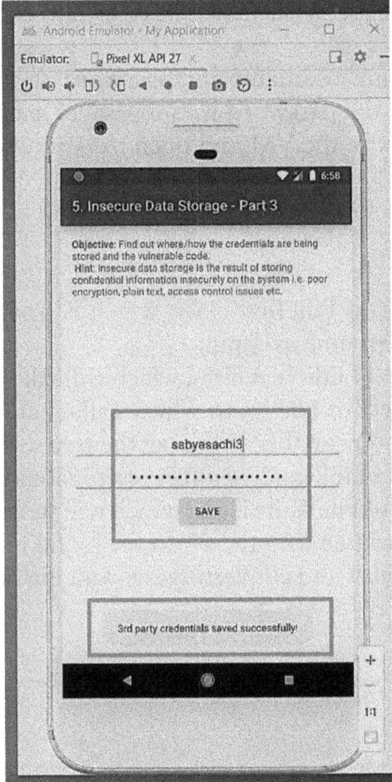

Figure 6.34 Giving some random username and password.

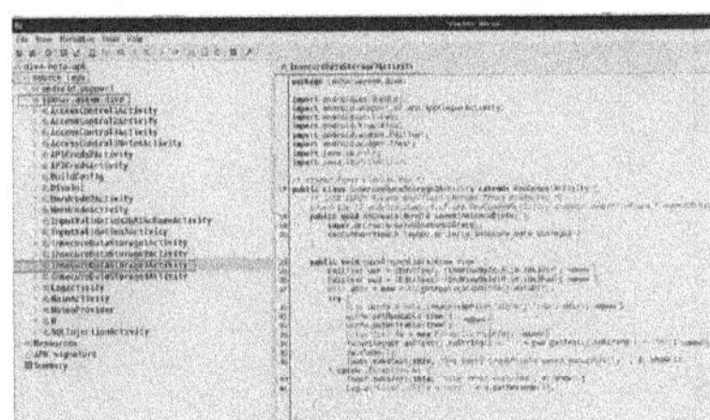

Figure 6.35 Source code of the challenge activity.

```
┌──(kali㉿kali)-[~]
└─$ adb -H 10.0.49.120 -P 5037 shell
generic_x86:/ # cd data/data/jakhar.aseem.diva/
generic_x86:/data/data/jakhar.aseem.diva # ls
app_textures app_webview cache code_cache databases files lib shared_prefs uinfo7738672600026999522tmp
generic_x86:/data/data/jakhar.aseem.diva # cat uinfo7738672600026999522tmp
sabyasachi3:sabyasachi_password3
generic_x86:/data/data/jakhar.aseem.diva #
```

Figure 6.36 Reading the file content of the created temporary file.

So, the objective is pretty clear. We have to find the source of our saved credentials.

Step #1: Provide some username and password. In this case, the username is "sabyasachi3" and the password is "sabyasachi_password3."

Step #2: Open Jadx in the VM and open the source code of the challenge activity

Now, for some the source code could be difficult to understand. Well, let me explain.

1. usr is the object of EditText class, which is basically a Text field that will take the username as input.
2. pwd is the object of EditText class, which will take the password as input.
3. uinfo is the object of File class, which will create a new temporary file named uinfo and probably will have the tmp as filename attached.
4. These are the readable and writable permissions given to the uinfo file.
5. fw is the object of FileWrite class, which will write the data into the file.
6. The data gets written into the uinfo file by taking the user input from usr and pwd object and converting the data into string using toString() method.

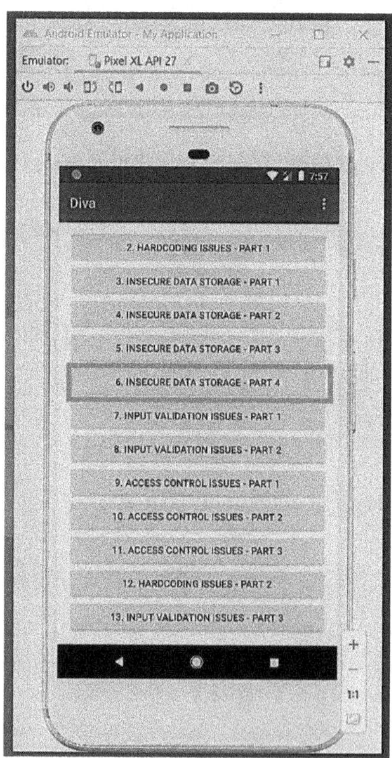

Figure 6.37 Challenge list.

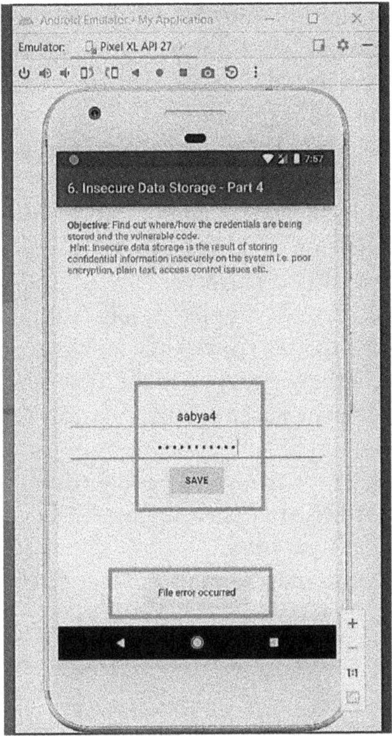

Figure 6.38 Giving some random username and password.

```
package jakhar.aseem.diva;

import android.os.Bundle;
import android.os.Environment;
import android.support.v7.app.AppCompatActivity;
import android.util.Log;
import android.view.View;
import android.widget.EditText;
import android.widget.Toast;
import java.io.File;
import java.io.FileWriter;

/* loaded from: classes.dex */
public class InsecureDataStorage4Activity extends AppCompatActivity {
    /* JADX INFO: Access modifiers changed from: protected */
    @Override // android.support.v7.app.AppCompatActivity, android.support.v4.app.FragmentAc
    public void onCreate(Bundle savedInstanceState) {
        super.onCreate(savedInstanceState);
        setContentView(R.layout.activity_insecure_data_storage4);
    }

    public void saveCredentials(View view) {
        EditText usr = (EditText) findViewById(R.id.ids4Usr);
        EditText pwd = (EditText) findViewById(R.id.ids4Pwd);
        File sdir = Environment.getExternalStorageDirectory();
        try {
            File uinfo = new File(sdir.getAbsolutePath() + "/.uinfo.txt");
            uinfo.setReadable(true);
            uinfo.setWritable(true);
            FileWriter fw = new FileWriter(uinfo);
            fw.write(usr.getText().toString() + ":" + pwd.getText().toString() + "\n");
            fw.close();
            Toast.makeText(this, "3rd party credentials saved successfully!", 0).show();
        } catch (Exception e) {
            Toast.makeText(this, "File error occurred", 0).show();
            Log.d("Diva", "File error: " + e.getMessage());
        }
    }
}
```

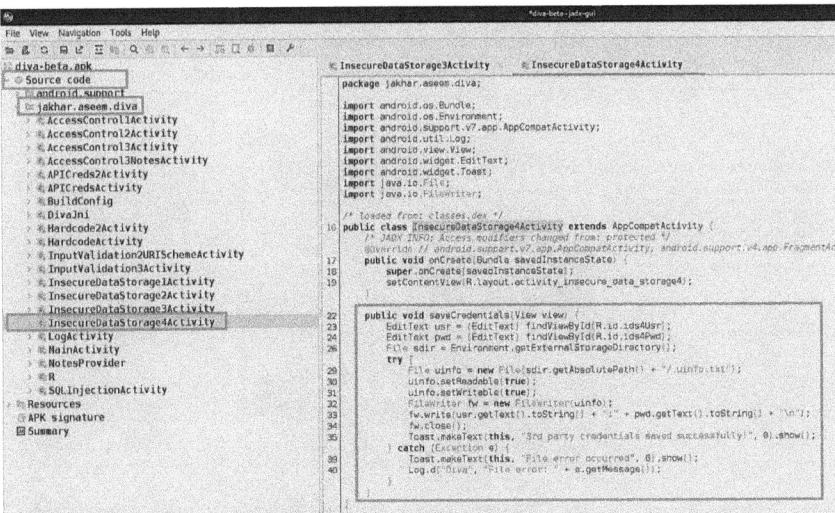

Figure 6.39 Source code of the challenge activity.

Now, our task is to locate the uinfo file. Hopefully, it will be inside the app package directory in the name of unifo or tmp as we have got two hints – the createTempFile() method and inside that tmp as a parameter that could be the directory/file name. Let's see!

Step #3: Open the terminal and make a connection and search for the uinfo file.

Well, the file is inside the app's package directory and with the name including both uinfo and tmp in it. Reading the file content gives the credentials in the username:password pattern.

Now, since this challenge is completed, let's jump into the next one. The next challenge is similar to this topic, that is, Insecure Data Storage – Part 4 but could be a bit challenging. As per the objective, we need to find the file or the location where the sensitive information is being stored. Let's see what files we get.

Step #1: Provide some username and password. In this case, the username is "sabya4" and the password is "sabya_pass4." This will not work, as the permission is not set. Let's proceed.

Step #2: Open Jadx and analyze the source code of the challenge activity.

Figure 6.40 Configuring app permissions.

Now, let me explain the code. In the saveCredential() method, there are two objects named usr and pwd of EditText class, which is used to take the user input of the username and password. Then, an object named uinfo is being created of class File that creates a new file named uinfo.txt preceding with a dot(.), which means that it will be a hidden file. Also between the try block and the pwd object line, that is, on line 26, the getExternalStorage-Directory() method has been invoked, which means we can guess that the file.uinfo.txt will be stored in the external storage device which could be SDCard. Let's find out.

Step #3: Go to the Android menu launcher and long hold the Diva app. Click on the App info.

Step #4: Select permission and turn on the Storage permission. If it's not done, the credentials won't be saved.

Step #5: Go to the app and now give the username and password. It's done.

Step #6: Go to the terminal and connect to the emulator. Go to the desired location.

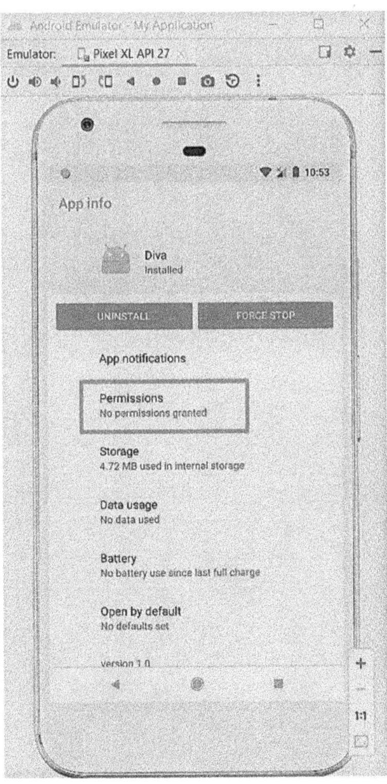

Figure 6.41 App info of the application.

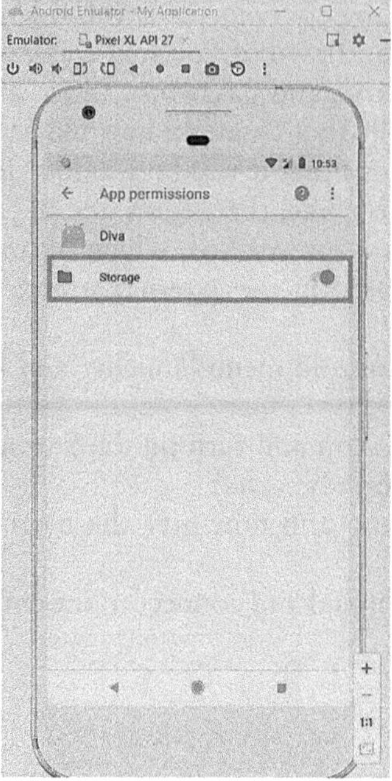

Figure 6.42 Turning on app permissions.

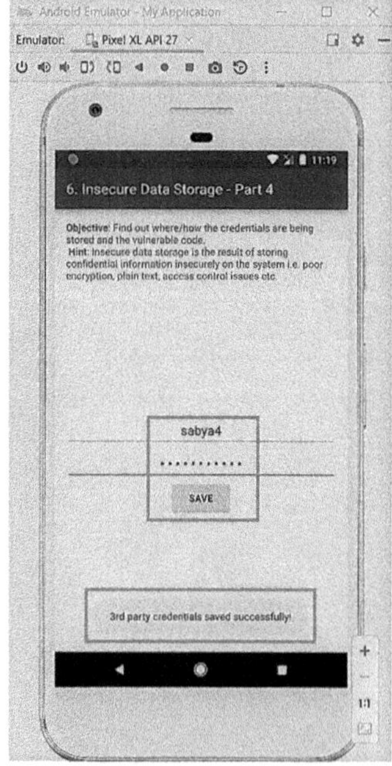

Figure 6.43 Giving random username and password. And it saved successfully.

```
File Actions Edit View Help

┌──(kali㉿kali)-[~]
└─$ adb -H 10.0.49.120 -P 5037 shell
generic_x86:/ # ls
acct       d            fstab.ranchu       init.rc                  oem    storage           var
bugreports data         fstab.ranchu.early init.usb.configfs.rc     proc   sys               vendor
cache      default.prop init               init.usb.rc              root   system
charger    dev          init.environ.rc    init.zygote32.rc         sbin   ueventd.ranchu.rc
config     etc          init.ranchu.rc     mnt                      sdcard ueventd.rc
generic_x86:/ # cd mnt/
generic_x86:/mnt # ls
appfuse asec expand media_rw obb runtime sdcard secure user
generic_x86:/mnt # cd sdcard/
generic_x86:/mnt/sdcard # ls
Alarms Android DCIM Download Movies Music Notifications Pictures Podcasts Ringtones
generic_x86:/mnt/sdcard # ls -la
total 52
drwxrwx--x 12 root sdcard_rw 4096 2023-01-27 23:19 .
drwx--x--x  4 root sdcard_rw 4096 2022-10-11 22:56 ..
-rw-rw----  1 root sdcard_rw   19 2023-01-27 23:19 .uinfo.txt
drwxrwx--x  2 root sdcard_rw 4096 2022-10-11 22:56 Alarms
drwxrwx--x  3 root sdcard_rw 4096 2022-10-11 22:56 Android
drwxrwx--x  2 root sdcard_rw 4096 2022-10-11 22:56 DCIM
drwxrwx--x  2 root sdcard_rw 4096 2022-12-20 11:00 Download
drwxrwx--x  2 root sdcard_rw 4096 2022-10-11 22:56 Movies
drwxrwx--x  2 root sdcard_rw 4096 2022-10-11 22:56 Music
drwxrwx--x  2 root sdcard_rw 4096 2022-10-11 22:56 Notifications
drwxrwx--x  2 root sdcard_rw 4096 2022-10-11 22:56 Pictures
drwxrwx--x  2 root sdcard_rw 4096 2022-10-11 22:56 Podcasts
drwxrwx--x  2 root sdcard_rw 4096 2022-10-11 22:56 Ringtones
generic_x86:/mnt/sdcard # cat .uinfo.txt
sabya4:sabya_pass4
generic_x86:/mnt/sdcard #
```

Figure 6.44 Exposed credentials.

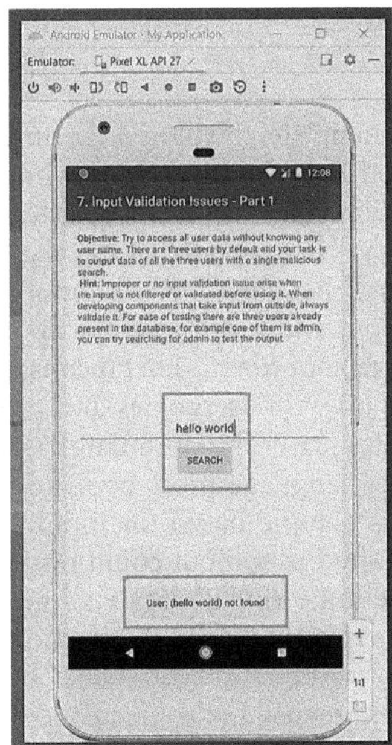

Figure 6.46 Supplying random user input.

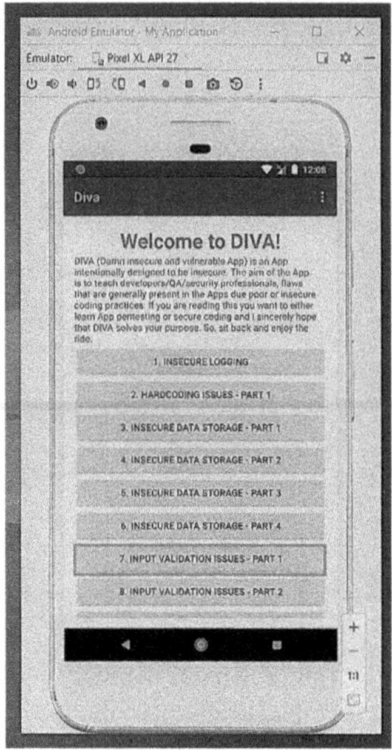

Figure 6.45 List of challenges.

Alright! So, we have completely explored the Insecure Data Storage Vulnerability. We understood how Android deals with data storage. Now let's explore the Input Validation vulnerabilities. So, let's go to the next one – Input Validation – Part 1. As per the objective of the challenge, we need to check whether the app is properly handling the user input or not. Let's find out.

Step #1: Give some input and check how the application responds.

It responds with (<user>) not found. That means, it could be an indication that it has something to do with usernames and probably those usernames will be stored somewhere, preferably in a database. And if it's a database, then some kind of query language must be in use. Second, the challenge name indicates that it is a topic based on Input Validation. That means, wrong or maliciously crafted user input could result in unwanted behavior. Therefore, there could be some kind of database injection vulnerability present. If the query language is SQL, then SQL Injection might be possible. Let's find out.

Step #2: Open Jadx and review the source code of the challenge activity.

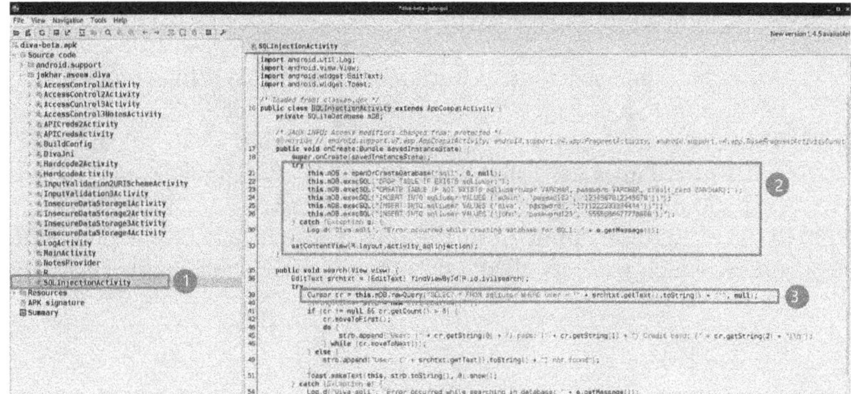

Figure 6.47 Source code of the challenge activity.

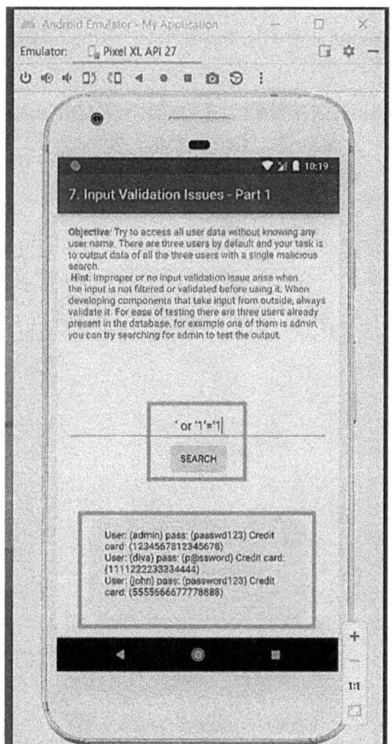

Figure 6.48 Exposed credentials.

Well, let me explain the code.

1 As we have guessed, there is a source file named SQLInjectionActivity, which somewhat depicts this activity's source code.
2 This onCreate() method creates a sqliuser table where all the hard-coded usernames are listed. Further, the query language it is using is SQL. So, there's a high chance of SQL Injection.
3 It takes the user input from the text field and passes it to the SQL query, which projects all the details of the particular username (if it exists).

```
public void search(View view) {
    EditText srchtxt = (EditText) findViewById(R.id.ivilsearch);
    try {
        Cursor cr = this.mDB.rawQuery("SELECT * FROM sqliuser WHERE user = '" + srchtxt.getText().toString() + "'", null);
        StringBuilder strb = new StringBuilder("");
        if (cr != null && cr.getCount() > 0) {
            cr.moveToFirst();
            do {
                strb.append("User: (" + cr.getString(0) + ") pass: (" + cr.getString(1) + ") Credit card: (" + cr.getString(2) + ")\n");
            } while (cr.moveToNext());
```

Figure 6.49 Source code where the malicious code has been injected.

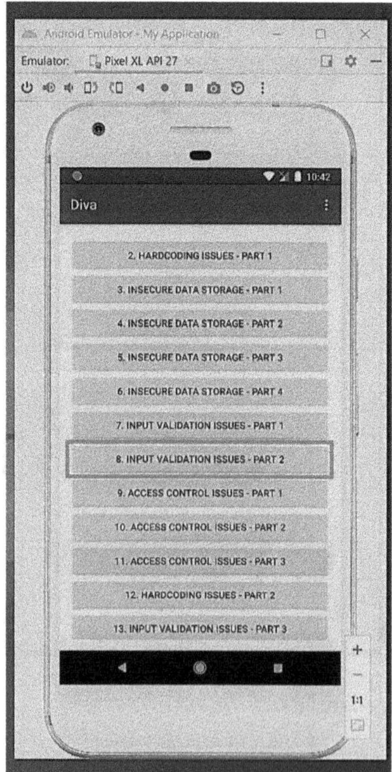

Figure 6.50 List of challenges.

So, let’s try SQL Injection and see if we could fetch all the user’s details including their credit card number.

For that, do the following in the emulator.

Step #3: Instead of username type: *‘ or ‘1’=’1* and click Search.

Well let me explain what is happening.

srchtxt.getText().toString() is taking the user input, converting it into string, and passing it to the query that is SELECT * FROM sqliuser WHERE user = ‘<fetched_username>’. Therefore, if we pass ‘ or ‘1’ = ‘1, then the query that is passed is SELECT * FROM sqliuser WHERE user = ‘’ or ‘1’=’1’. That means SQL is getting a condition that if username is ‘’, that is, empty and if it exists then get the details or if the condition is true which is 1=1 which is universally true, then show all the details of all the users since there was * wildcard present in the SQL Query.

Note: If you are having difficulty understanding it, it is recommended to have a look at SQL first.

Well, this challenge was interesting. Let’s jump to the next one, that is, Input Validation – Part 2. And as per the challenge, objective seems like it is

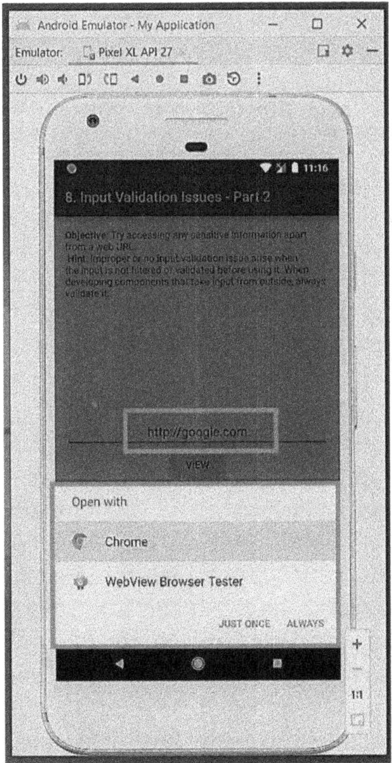

Figure 6.51 Supplying web URL.

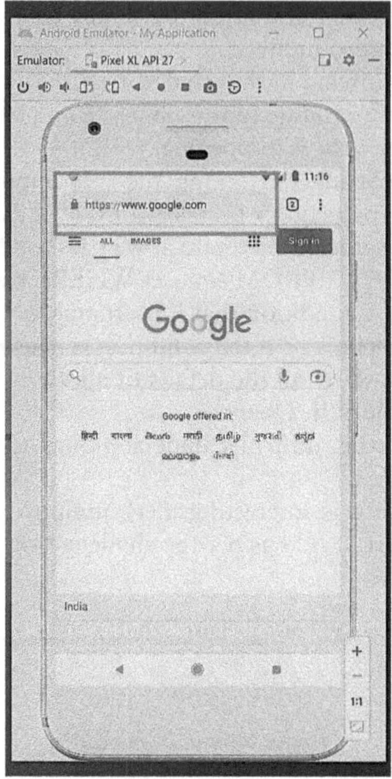

Figure 6.52 URL opens in the browser.

similar to the previous one with a bit higher difficulty level. It asks to give input a URI. So, let's pass a web URL and see what happens.

Step #1: Supplying web URL in the text field. For example, http://google.com

Step #2: Checking the source code for some clue.

Well, the source code might look scary, but it isn't. Let me explain.

At first, we are importing some webkit packages. Webkit is a rendering library that renders web pages. So, the URL we provide, webkit renders it and shows it to us. Second, the onCreate() method creates a WebView, which basically allows webpages to display as a part of your activity. In addition, the method makes sure that the JavaScript is enabled; else fewer websites that are using JavaScript might not function properly. Lastly, the get() method takes the web URL as the user input from the text field named as urlText, which is an object of EditText class. Then, the wview, which is an object of the WebView class, loads the URL by fetching the user input from the urlText field and converting into string.

Pretty straightforward right?

```
*diva-beta - jadx-gui
File  View  Navigation  Tools  Help

diva-beta.apk
  Source code
    android.support
    jakhar.aseem.diva
      AccessControl1Activity
      AccessControl2Activity
      AccessControl3Activity
      AccessControl3NotesActivity
      APICreds2Activity
      APICredsActivity
      BuildConfig
      DivaJni
      Hardcode2Activity
      HardcodeActivity
      InputValidation2URISchemeActivity
      InputValidation3Activity
      InsecureDataStorage1Activity
      InsecureDataStorage2Activity
      InsecureDataStorage3Activity
      InsecureDataStorage4Activity
      LogActivity
      MainActivity
      NotesProvider
      R
      SQLInjectionActivity
  Resources
  APK signature
  Summary

SQLInjectionActivity    InputValidation2URISchemeActivity

package jakhar.aseem.diva;

import android.os.Bundle;
import android.support.v7.app.AppCompatActivity;
import android.view.View;
import android.webkit.WebSettings;
import android.webkit.WebView;
import android.widget.EditText;

/* loaded from: classes.dex */
public class InputValidation2URISchemeActivity extends AppCompatActivity {
    /* JADX INFO: Access modifiers changed from: protected */
    @Override // android.support.v7.app.AppCompatActivity, android.support.v4
    public void onCreate(Bundle savedInstanceState) {
        super.onCreate(savedInstanceState);
        setContentView(R.layout.activity_input_validation2_urischeme);
        WebView wview = (WebView) findViewById(R.id.ivi2wview);
        WebSettings wset = wview.getSettings();
        wset.setJavaScriptEnabled(true);
    }

    public void get(View view) {
        EditText uriText = (EditText) findViewById(R.id.ivi2uri);
        WebView wview = (WebView) findViewById(R.id.ivi2wview);
        wview.loadUrl(uriText.getText().toString());
    }
}
```

Figure 6.53 Source code of the challenge activity.

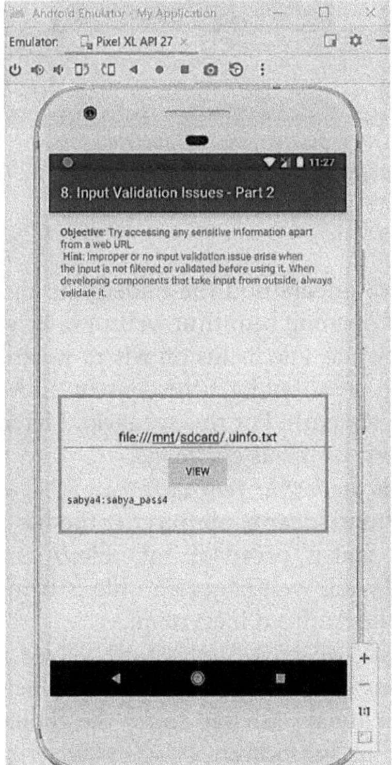

Figure 6.54 The file has been listed.

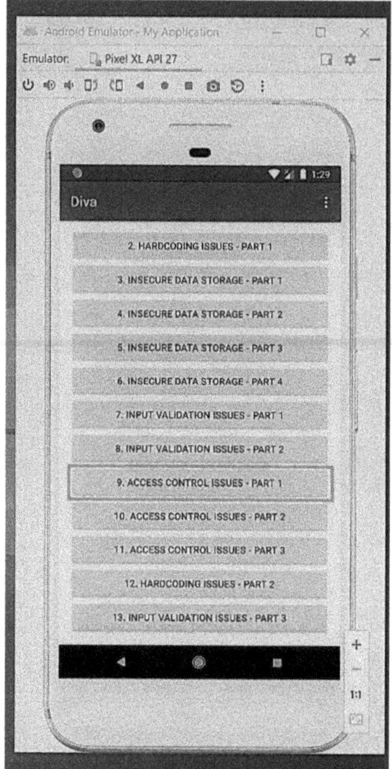

Figure 6.55 List of challenges.

Now, since we have understood the underlying mechanism of the challenge, then instead of opening beautiful websites, we will try something different preferably known as File inclusion where we are going to provide the file path and in return we should get the content of the file.

Step #3: Supplying file path. For this scenario, I have supplied the credential file we found in our previous challenge.

The actual syntax is *file://<file_path>*.

Therefore, – file://**/mnt/sdcard/.uinfo.txt** is the file path and the **file://** is the File URI Scheme, that is, protocol. Take **file://** similar to **http://**, where **http** is used for displaying web pages and **file** is being used for displaying local files present in the Android filesystem.

Well that being said, we have successfully solved this challenge, which was fun and interesting too. Now, let's jump onto the next challenge.

Remember, you are always advised to try the challenges by yourself first and then referring to the solutions.

So, the next challenge states Access Control Issues – Part 1 and by the name it seems like we have to deal with some kind of authorization. Well,

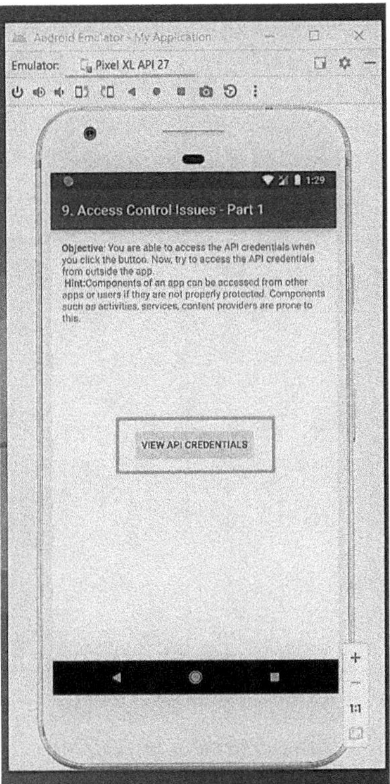

Figure 6.56 The challenge activity.

the objective of this challenge states that the API creds can be viewed whenever we are interacting with the app. But our motive is to fire up the activity outside the app. So, let's see what we need to do.

Step #1: Click on View API Credentials to see what actions are happening.

So, upon clicking the button, an activity is launched

Our target is to launch the same API Creds activity outside of the app. So for that, we need to grab the intent and fire it up. Let's see where is it.

Step #2: Review the source code of the challenge activity.

Well, the code might seem to be confusing, but wait hold on, let me explain.

At first in line 20, i is the object of the Intent class. That means, a new Intent has been created. In line 21, the Intent is given, which is in double quotes (""). And in line 24, the startActivity() method starts an activity based upon the given intent specified in line 21.

So, if we could specifically grab the intent and fire it up, then the activity might start. Let's see!

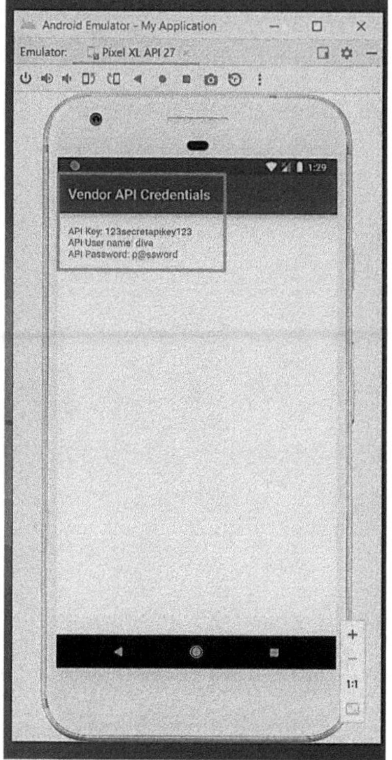

Figure 6.57 The API Cred activity.

```
package jakhar.aseem.diva;

import android.content.Intent;
import android.os.Bundle;
import android.support.v7.app.AppCompatActivity;
import android.util.Log;
import android.view.View;
import android.widget.Toast;

/* loaded from: classes.dex */
public class AccessControl1Activity extends AppCompatActivity {
    /* JADX INFO: Access modifiers changed from: protected */
    @Override // android.support.v7.app.AppCompatActivity, android.support.v4.app.Frag
    public void onCreate(Bundle savedInstanceState) {
        super.onCreate(savedInstanceState);
        setContentView(R.layout.activity_access_control1);
    }

    public void viewAPICredentials(View view) {
        Intent i = new Intent();
        i.setAction("jakhar.aseem.diva.action.VIEW_CREDS");
        if (i.resolveActivity(getPackageManager()) != null) {
            startActivity(i);
            return;
        }
        Toast.makeText(this, "Error while getting API details", 0).show();
        Log.e("Diva-aci1", "Couldn't resolve the Intent VIEW_CREDS to our activity");
    }
}
```

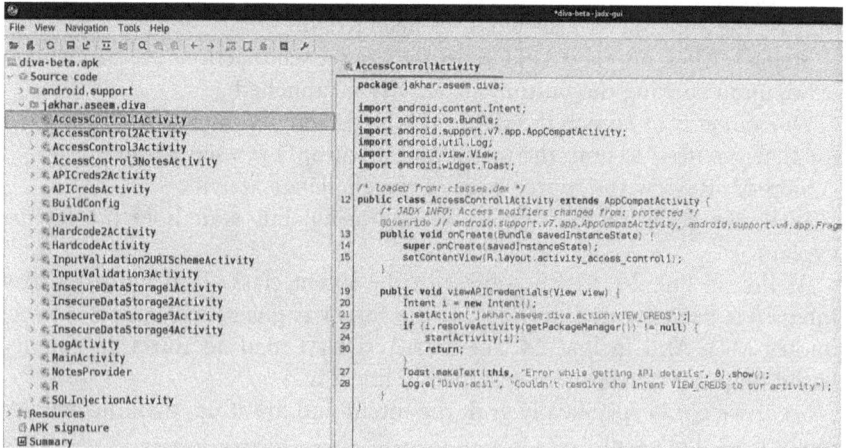

Figure 6.58 Source code of the challenge activity.

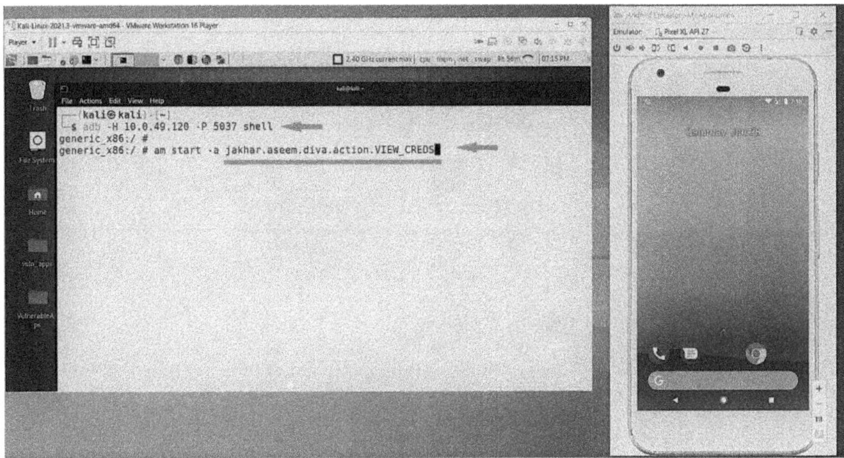

Figure 6.59 Firing up the activity separately.

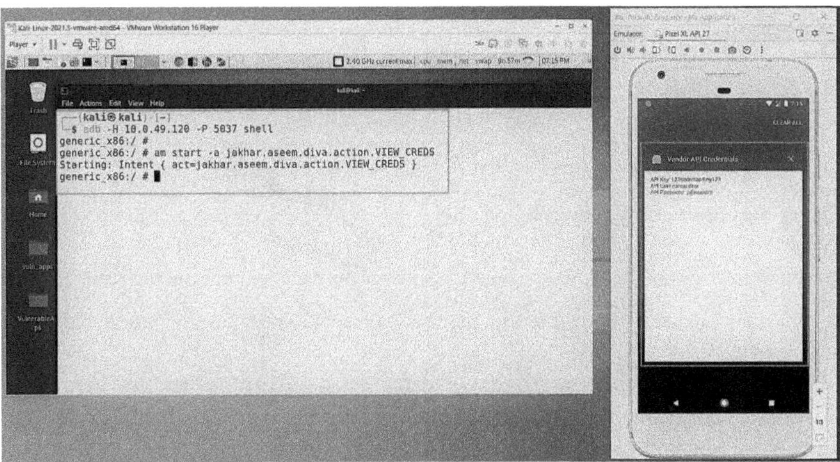

Figure 6.60 The activity launches.

Step #3: Grab the intent, open terminal and fire it up separately. Make sure the app is closed and the shell of the emulator is opened in the VM. Type: am start -a <activity_name>.

Here, **am** stands for Activity Manager, **start** means to start the activity, and **-a** is the switch after which we pass the activity name.

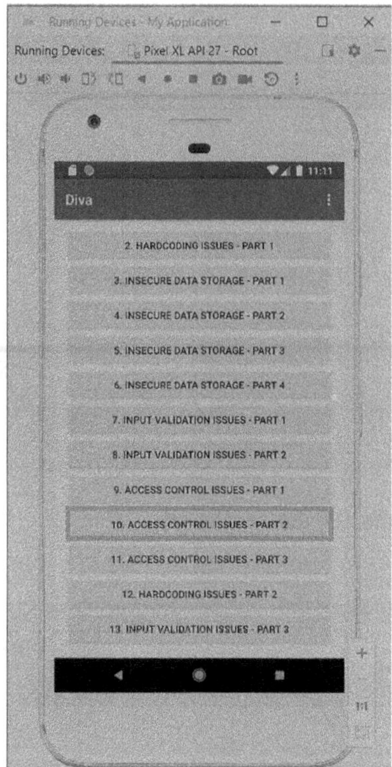

Figure 6.61 List of challenges.

So, if the steps bounced off your head, then no worries. Let me summarize it for you.

Basically, the window that we refer to as Activity in the jargon of Android programming will only appear if we hit the View API Credentials button when we are in the app. This implies that we should only see the API Creds when interacting with the app. Now, if the Intent is poorly configured, then we should be able to call the activity outside the app as well. So, using the source code, we analyzed the activity name, which is responsible for firing up the activity once we click the button. So, we grabbed it, and using Activity Manager (am in command), we separately fired it up. Since the app was poorly built, we get to see the activity launching outside the app also.

Alright! So, we also succeeded in resolving this problem. It had a lot of learning potential and was pretty intriguing.

Next, we've got Access Control Issues – Part 2. As the name goes, we again have to bypass the access controls and fetch an unauthorized content. So, as far as the problem description goes that we have to fetch the Tveeter API Creds without giving the pin. So, let's see how can we exploit this.

So, the challenge looks like this.

So, upon clicking the "VIEW API TVEETER CREDENTIALS" button, we get this activity.

So, our job is to view this activity outside of the app. So, like the previous challenge let's open Jadx and see for the code.

Step #1: Open Jadx to view the source code of the app and find the activity.

So, we got two activities that point to the same challenge. Upon seeing the first activity, that is, Access Control 2 Activity, we get to see the activity that is responsible in popping up the API Creds page.

So, like the previous one, we will be using the activity manager to explicitly pop up the activity.

Step #2: Go to the terminal and type – am start -a jakhar.aseem.diva.action.VIEWCREDS2.

Note that I am issuing the command inside the shell of the Android emulator.

But it seems like it's asking for password.

Step #3: Reviewing the second activity APICreds2Activity.

So, from the code, we can comprehend that to view the API Creds, we need to give some pin, which is being validated by the variable named "chk_pin." Also, in the activity named "AccessControl2Activity," we can see that the same variable "chk_pin" is getting checked.

Step #4: Find the variable chk_pin and find its purpose.

So, now we need to search for the variable chk_pin. So, in the "AccessControl2Activity," we can see that the whole access path is "R.strings.chk_pin."

So, there's a possibility that chk_pin variable is located inside the strings.xml. So, to get the strings.xml its inside, Resouces/ > resources.arsc > res/ > values/ > strings.xml

So, if we observe the strings.xml file, we get to see the chk_pin.

Here, we get to see <string name="chk_pin">check_pin</string>. So, the chk_pin is just an alias for the actual variable check_pin.

Now, we can try bypassing it by issuing the command using activity manager. But before that, if we check the AndoidManifest.xml file, then we can observe that under the APICreds2Activity, there is an intent filter on VIEW_CREDS2. So, this is what we have to take care of while issuing the command in the terminal.

Step #5: Go to the terminal and type: am start -n jakhar.aseem.diva/.APICreds2Activity -a jakhar.aseem.diva/.VIEW_CREDS2 –ez check_pin false.

So, if the command is to be explained, it would be like: we are asking the activity manager to start an activity (represented by -a) name

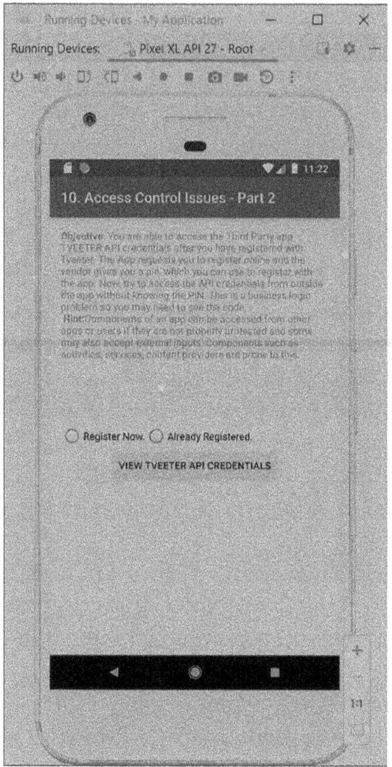

Figure 6.62 The challenge activity.

VIEW_CREDS2 and it is to be shown by dumping the activity (represented by -n) name APICreds2Activity. Also, use an extra variable (represented as e) whose datatype would be boolean (represented by z) named check_pin and keep its value as false (represented as –ez check_pin false).

So, in summary in order to directly fetch VIEW_CREDS, we need to give a pin; so, providing its value as false to get the VIEW_CREDS, we need to fetch it via APICreds2Activity. So, we dumped APICreds2Activity in order to fetch VIEW_CREDS.

Well, it was a bit intimidating, especially the exploitation part but we learnt a lot.

Let's now jump to the next challenge named Access Control Issues – Part 3.

So, as per the description, we need to access the private notes without giving the secret pin. So, let's see how can we exploit it. Also, before exploiting, let's first explore the challenge. So here it asks to create or enter a pin. So, let's give 1234, click on the create pin button, then click on the Go to Private notes button, then give the PIN, and then click on Access Private notes and the notes should be visible.

Figure 6.63 API Creds.

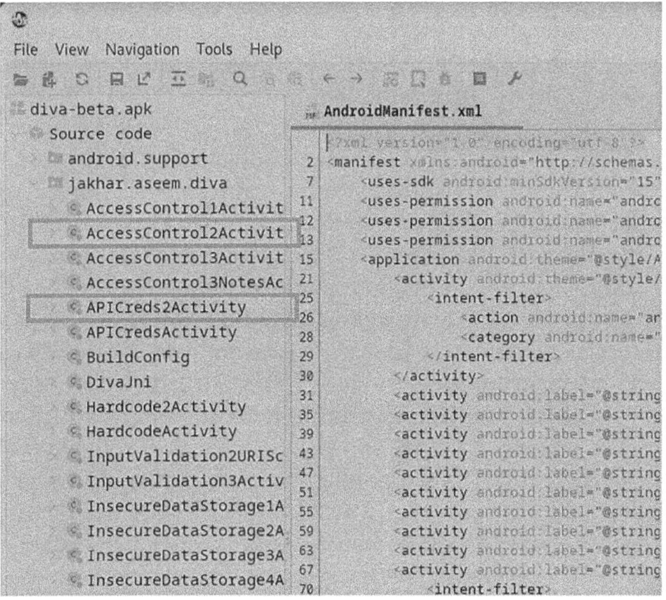

Figure 6.64 Jadx window.

Step #1: Open Jadx and search for the related activity in AndroidManifest.xml

So, these are the activities that are associated with this challenge.

Step #2: Review the activity codes.

First, we'll explore the AccessControl3Activity.

So, to explain the code:

The provided code appears to be a method named addPin() within a class.

SharedPreferences spref = PreferenceManager.getDefaultSharedPreferences(this); – This line retrieves the default SharedPreferences instance using PreferenceManager in the current context (this). SharedPreferences is an interface that allows you to store and retrieve key-value pairs persistently.

SharedPreferences.Editor spedit = spref.edit(); – This line obtains an editor object from the SharedPreferences instance. The editor allows you to make changes to the shared preferences.

EditText pinTxt = (EditText) findViewById(R.id.aci3Pin); – This line finds the EditText view with the ID aci3Pin in the current layout and assigns it to the pinTxt variable. The findViewById method is used to retrieve a reference to a view based on its ID.

String pin = pinTxt.getText().toString(); – This line retrieves the text entered in the pinTxt EditText view and converts it to a string by calling the toString method.

if (pin == null || pin.isEmpty()) {. . .} – This line checks if the pin variable is either null or an empty string. If it is, a Toast message is displayed, indicating that a valid PIN should be entered, and the method returns, exiting further execution.

Button vbutton = (Button) findViewById(R.id.aci3viewbutton); – This line finds the Button view with the ID aci3viewbutton in the current layout and assigns it to the vbutton variable.

spedit.putString(getString(R.string.pkey), pin); – This line puts a string value (pin) into the SharedPreferences.Editor object (spedit) using the key obtained from the string resource with the name pkey. The getString method retrieves the actual string value associated with the resource ID.

spedit.commit(); – This line saves the changes made to the shared preferences by calling the commit method on the SharedPreferences.Editor object. This ensures that the new PIN value is stored persistently.

if (vbutton.getVisibility()!= 0) {. . .} – This line checks if the visibility of the vbutton is not currently set to View.VISIBLE (0), meaning it's not visible. If that's the case, the visibility of vbutton is set to View.VISIBLE, making it visible.

Toast.makeText(this, "PIN Created successfully. Private notes are now protected with PIN", 0).show(); This line displays a short-duration Toast message with the text "PIN Created successfully. Private notes are now protected with PIN" to provide feedback to the user about the successful creation of the PIN.

```java
AndroidManifest.xml    AccessControl2Activity
package jakhar.aseem.diva;

import android.content.Intent;
import android.os.Bundle;
import android.support.v7.app.AppCompatActivity;
import android.util.Log;
import android.view.View;
import android.widget.RadioButton;
import android.widget.Toast;

/* loaded from: classes.dex */
public class AccessControl2Activity extends AppCompatActivity {
    /* JADX INFO: Access modifiers changed from: protected */
    @Override // android.support.v7.app.AppCompatActivity, android.support.v4.app.FragmentActivity, android.support.v4.app.BaseFr
    public void onCreate(Bundle savedInstanceState) {
        super.onCreate(savedInstanceState);
        setContentView(R.layout.activity_access_control2);
    }

    public void viewAPICredentials(View view) {
        RadioButton rbregnow = (RadioButton) findViewById(R.id.aci2rbregnow);
        Intent i = new Intent();
        boolean chk_pin = rbregnow.isChecked();
        i.setAction("jakhar.aseem.diva.action.VIEW_CREDS2");
        i.putExtra(getString(R.string.chk_pin), chk_pin);
        if (i.resolveActivity(getPackageManager()) != null) {
            startActivity(i);
            return;
        }
        Toast.makeText(this, "Error while getting Tveeter API details", 0).show();
        Log.e("Diva-aci1", "Couldn't resolve the Intent VIEW_CREDS2 to our activity");
    }
}
```

Figure 6.65 Access Control 2 Activity code window.

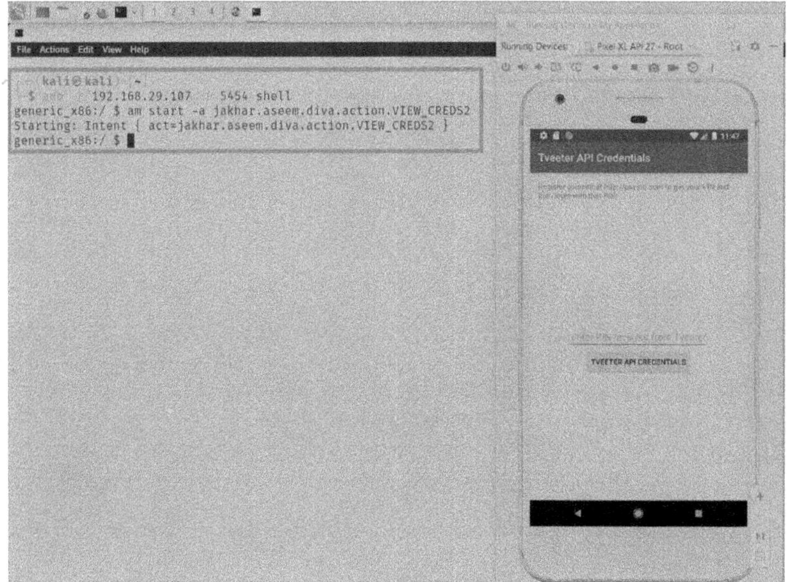

Figure 6.66 Issuing command in the Android shell to access the API Creds activity.

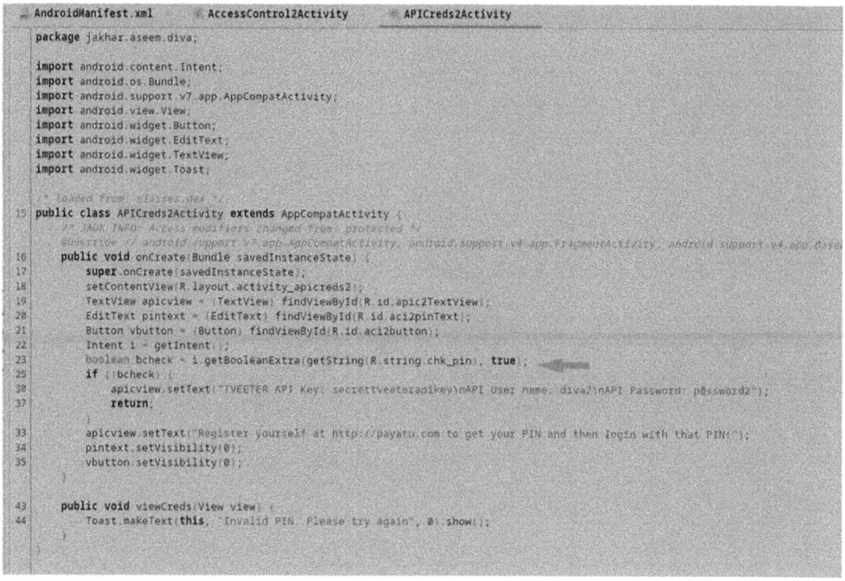

```
AndroidManifest.xml    AccessControl2Activity    APICreds2Activity
package jakhar.aseem.diva;

import android.content.Intent;
import android.os.Bundle;
import android.support.v7.app.AppCompatActivity;
import android.view.View;
import android.widget.Button;
import android.widget.EditText;
import android.widget.TextView;
import android.widget.Toast;

/* loaded from: classes.dex */
public class APICreds2Activity extends AppCompatActivity {
    /* JADX INFO: Access modifiers changed from: protected */
    @Override // android.support.v7.app.AppCompatActivity, android.support.v4.app.FragmentActivity, android.support.v4.app.Base
    public void onCreate(Bundle savedInstanceState) {
        super.onCreate(savedInstanceState);
        setContentView(R.layout.activity_apicreds2);
        TextView apicview = (TextView) findViewById(R.id.apic2TextView);
        EditText pintext = (EditText) findViewById(R.id.aci2pinText);
        Button vbutton = (Button) findViewById(R.id.aci2button);
        Intent i = getIntent();
        boolean bcheck = i.getBooleanExtra(getString(R.string.chk_pin), true);
        if (!bcheck) {
            apicview.setText("TVEETER API Key: secrettveeterapikey\nAPI User name: diva2\nAPI Password: p@ssword2");
            return;
        }
        apicview.setText("Register yourself at http://payatu.com to get your PIN and then login with that PIN!");
        pintext.setVisibility(0);
        vbutton.setVisibility(0);
    }

    public void viewCreds(View view) {
        Toast.makeText(this, "Invalid PIN. Please try again", 0).show();
    }
}
```

Figure 6.67 APICreds2Activity source code window.

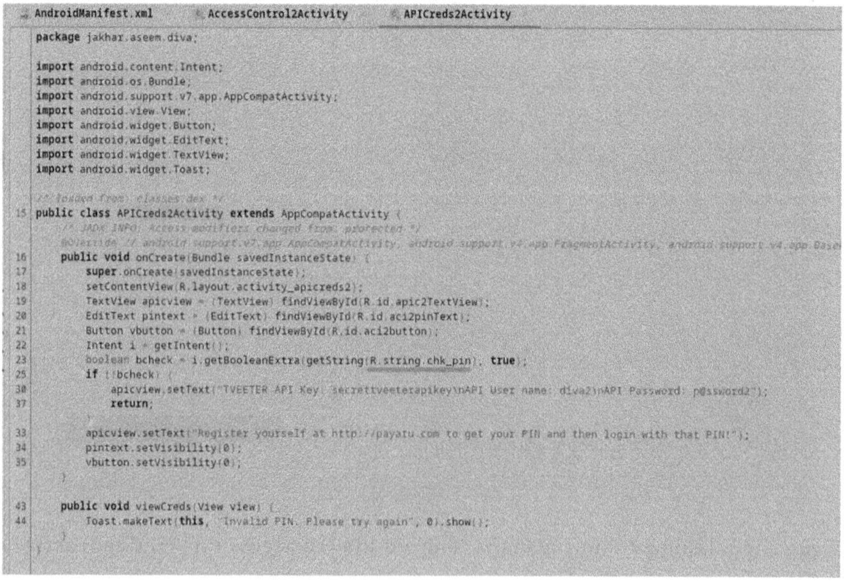

```
AndroidManifest.xml    AccessControl2Activity    APICreds2Activity
package jakhar.aseem.diva;

import android.content.Intent;
import android.os.Bundle;
import android.support.v7.app.AppCompatActivity;
import android.view.View;
import android.widget.Button;
import android.widget.EditText;
import android.widget.TextView;
import android.widget.Toast;

/* loaded from: classes.dex */
public class APICreds2Activity extends AppCompatActivity {
    /* JADX INFO: Access modifiers changed from: protected */
    @Override // android.support.v7.app.AppCompatActivity, android.support.v4.app.FragmentActivity, android.support.v4.app.Base
    public void onCreate(Bundle savedInstanceState) {
        super.onCreate(savedInstanceState);
        setContentView(R.layout.activity_apicreds2);
        TextView apicview = (TextView) findViewById(R.id.apic2TextView);
        EditText pintext = (EditText) findViewById(R.id.aci2pinText);
        Button vbutton = (Button) findViewById(R.id.aci2button);
        Intent i = getIntent();
        boolean bcheck = i.getBooleanExtra(getString(R.string.chk_pin), true);
        if (!bcheck) {
            apicview.setText("TVEETER API Key: secrettveeterapikey\nAPI User name: diva2\nAPI Password: p@ssword2");
            return;
        }
        apicview.setText("Register yourself at http://payatu.com to get your PIN and then login with that PIN!");
        pintext.setVisibility(0);
        vbutton.setVisibility(0);
    }

    public void viewCreds(View view) {
        Toast.makeText(this, "Invalid PIN. Please try again", 0).show();
    }
}
```


Figure 6.68 Finding chk_pin variable.

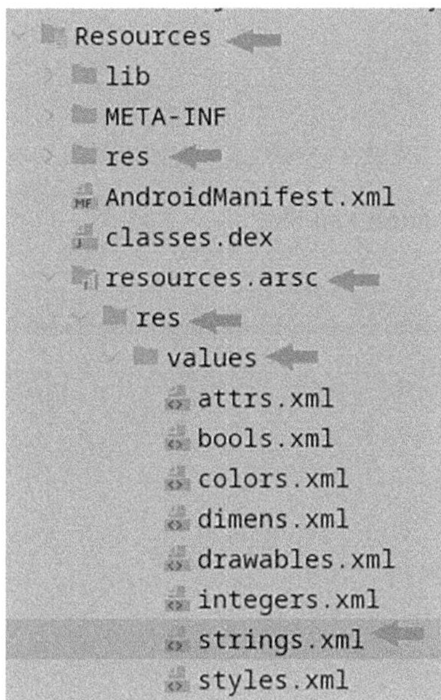

Figure 6.69 Path for strings.xml.

```
AndroidManifest.xml    AccessControl2Activity    APICreds2Activity    R    res/values/strings.xml
<string name="abc_searchview_description_search">Search</string>
<string name="abc_searchview_description_submit">Submit query</string>
<string name="abc_searchview_description_voice">Voice search</string>
<string name="abc_shareactionprovider_share_with">Share with</string>
<string name="abc_shareactionprovider_share_with_application">Share with %s</string>
<string name="abc_toolbar_collapse_description">Collapse</string>
<string name="status_bar_notification_info_overflow">999+</string>
<string name="aci1_intro">Objective: You are able to access the API credentials when you click the button. Now, try to
<string name="aci1_view">VIEW API CREDENTIALS</string>
<string name="aci2_alreadyreg">Already Registered.</string>
<string name="aci2_intro">Objective: You are able to access the Third Party app TVEETER API credentials after you have
<string name="aci2_regnow">Register Now.</string>
<string name="aci2_view">VIEW TVEETER API CREDENTIALS</string>
<string name="aci3_accessnotes">ACCESS PRIVATE NOTES</string>
<string name="aci3_createpin">CREATE/CHANGE PIN</string>
<string name="aci3_enterpin">Enter 4 Digit PIN</string>
<string name="aci3_gotonotes">GO TO PRIVATE NOTES</string>
<string name="aci3_intro">Objective: This is a private notes application. You can create a PIN once and access your not
<string name="action_settings">Settings</string>
<string name="apic2_enter">Enter PIN received from Tveeter</string>
<string name="apic2_label">Tveeter API Credentials</string>
<string name="apic_label">Vendor API Credentials</string>
<string name="app_name">Diva</string>
<string name="appbar_scrolling_view_behavior">android.support.design.widget.AppBarLayout$ScrollingViewBehavior</string>
<string name="character_counter_pattern">%1$d/%2$d</string>
<string name="chk_pin">check_pin</string>
<string name="d1">1. Insecure Logging</string>
<string name="d10">10. Access Control Issues - Part 2</string>
<string name="d11">11. Access Control Issues - Part 3</string>
<string name="d12">12. Hardcoding Issues - Part 2</string>
<string name="d13">13. Input Validation Issues - Part 3</string>
```

Figure 6.70 The chk_pin variable in strings.xml.

```
<activity android:label="@string/d10" android:name="jakhar.aseem.diva.AccessControl2Activity"/>
<activity android:label="@string/apic2_label" android:name="jakhar.aseem.diva.APICreds2Activity">
    <intent-filter>
        <action android:name="jakhar.aseem.diva.action.VIEW_CREDS2"/>
        <category android:name="android.intent.category.DEFAULT"/>
    </intent-filter>
</activity>
```

Figure 6.71 AndroidManifest.xml file.

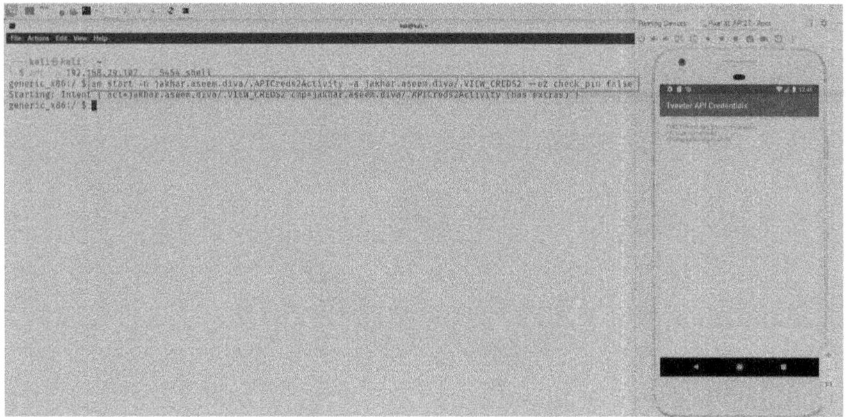

Figure 6.72 The API Creds activity popped up.

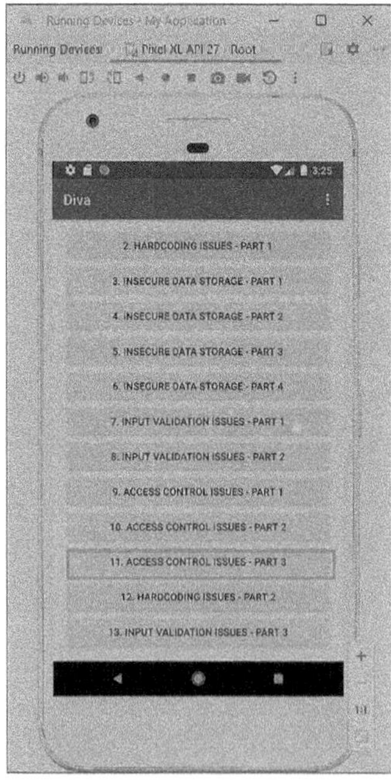

Figure 6.73 List of challenges.

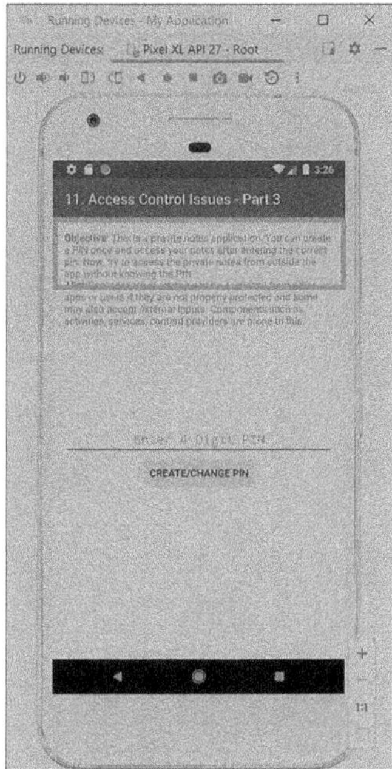

Figure 6.74 Challenge description.

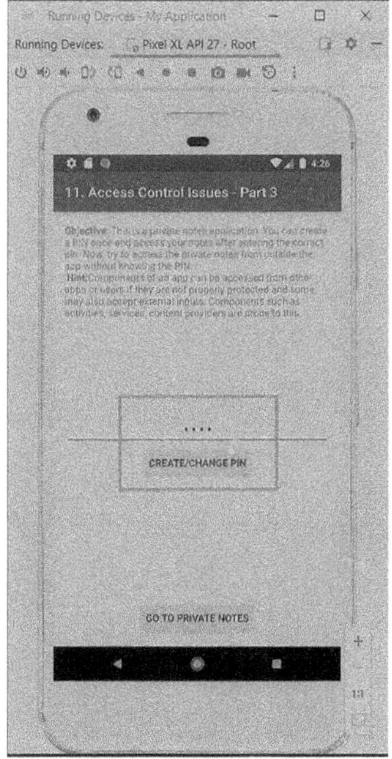

Figure 6.75 Create a PIN and click on the "Create/Change PIN" button.

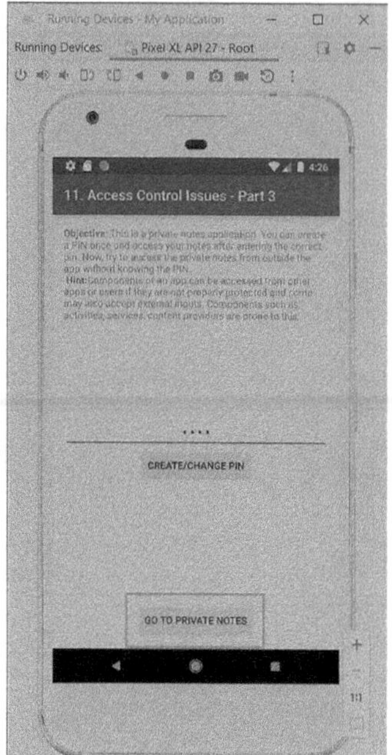

Figure 6.76 Click on Go to Private Notes.

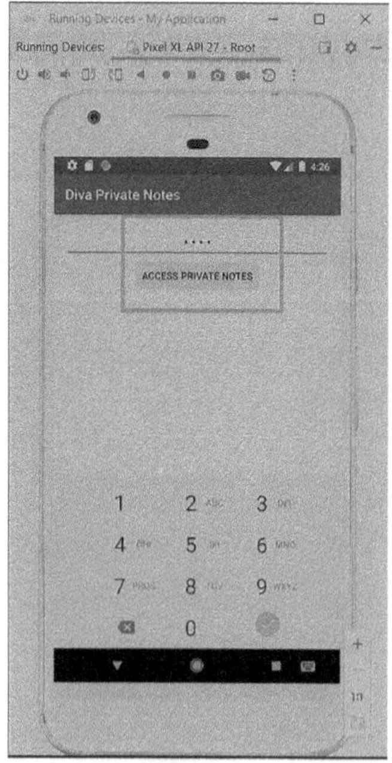

Figure 6.77 Enter the PIN and click on the Access Private Notes button.

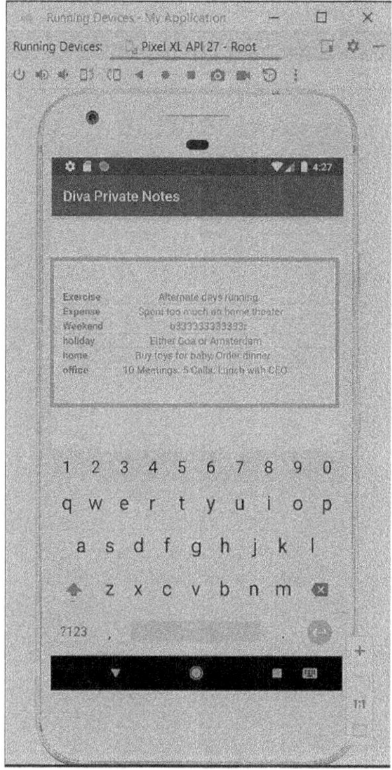

Figure 6.78 The Private Notes window.

Figure 6.79 Locating activities in AndroidManifest.xml.

In summary, the addPin() method is responsible for handling the creation of a PIN by retrieving the entered PIN from an EditText view, storing it in the shared preferences using a specific key, and displaying a Toast message to inform the user about the outcome of the operation.

So, we kind of can comprehend that the PIN is being saved into Shared Preferences.

Next let's look into the AccessControl3NotesActivity.

So, to explain you the code, the code represents a class named AccessControl3NotesActivity that extends the AppCompatActivity class. This class likely corresponds to an activity within an Android application.

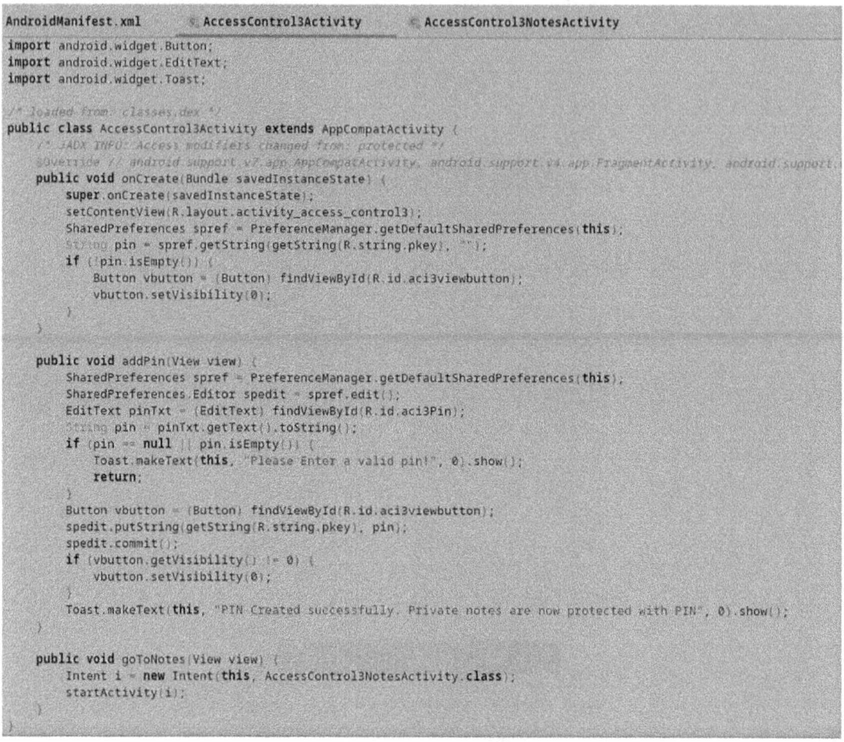

```
AndroidManifest.xml    AccessControl3Activity    AccessControl3NotesActivity
import android.widget.Button;
import android.widget.EditText;
import android.widget.Toast;

/* loaded from: classes.dex */
public class AccessControl3Activity extends AppCompatActivity {
    /* JADX INFO: Access modifiers changed from: protected */
    @Override // android.support.v7.app.AppCompatActivity, android.support.v4.app.FragmentActivity, android.support.
    public void onCreate(Bundle savedInstanceState) {
        super.onCreate(savedInstanceState);
        setContentView(R.layout.activity_access_control3);
        SharedPreferences spref = PreferenceManager.getDefaultSharedPreferences(this);
        String pin = spref.getString(getString(R.string.pkey), "");
        if (!pin.isEmpty()) {
            Button vbutton = (Button) findViewById(R.id.aci3viewbutton);
            vbutton.setVisibility(0);
        }
    }

    public void addPin(View view) {
        SharedPreferences spref = PreferenceManager.getDefaultSharedPreferences(this);
        SharedPreferences.Editor spedit = spref.edit();
        EditText pinTxt = (EditText) findViewById(R.id.aci3Pin);
        String pin = pinTxt.getText().toString();
        if (pin == null || pin.isEmpty()) {
            Toast.makeText(this, "Please Enter a valid pin!", 0).show();
            return;
        }
        Button vbutton = (Button) findViewById(R.id.aci3viewbutton);
        spedit.putString(getString(R.string.pkey), pin);
        spedit.commit();
        if (vbutton.getVisibility() != 0) {
            vbutton.setVisibility(0);
        }
        Toast.makeText(this, "PIN Created successfully. Private notes are now protected with PIN", 0).show();
    }

    public void goToNotes(View view) {
        Intent i = new Intent(this, AccessControl3NotesActivity.class);
        startActivity(i);
    }
}
```

Figure 6.80 AccessControl3Activity source code.

@Override This annotation indicates that the following method is intended to override a method from the superclass.

protected void onCreate(Bundle savedInstanceState) – This method is an overridden implementation of the onCreate method, which is a lifecycle callback method called when the activity is being created. It is responsible for initializing the activity and setting its content view by inflating the layout resource file R.layout.activity_access_control3_notes.

public void accessNotes(View view) – This method is a public member function that is likely associated with a button click event. It takes a View parameter, which represents the clicked view. This method is invoked when the associated button is clicked.

EditText pinTxt = (EditText) findViewById(R.id.aci3notesPinText); – This line retrieves the EditText view with the ID aci3notesPinText from the current layout and assigns it to the pinTxt variable.

Button abutton = (Button) findViewById(R.id.aci3naccessbutton); – This line retrieves the Button view with the ID aci3naccessbutton from the current layout and assigns it to the abutton variable.

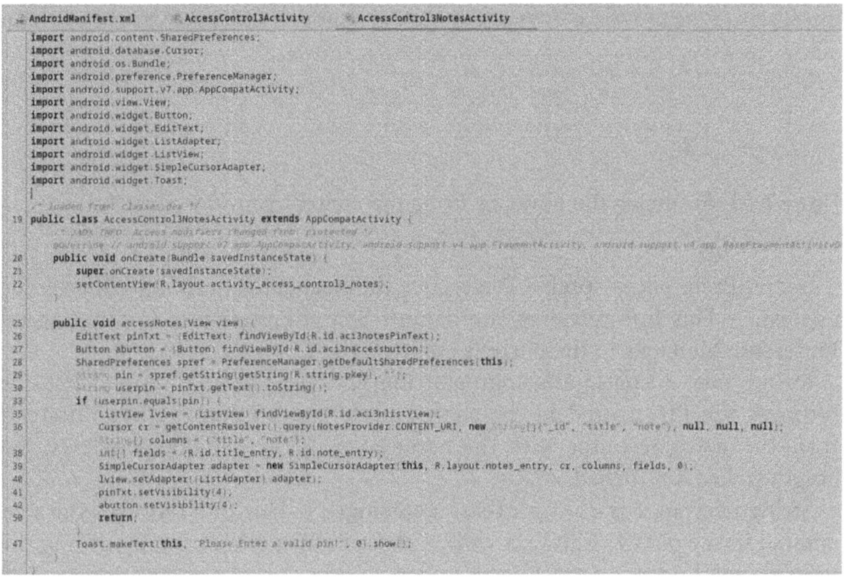

```
AndroidManifest.xml    AccessControl3Activity    AccessControl3NotesActivity
import android.content.SharedPreferences;
import android.database.Cursor;
import android.os.Bundle;
import android.preference.PreferenceManager;
import android.support.v7.app.AppCompatActivity;
import android.view.View;
import android.widget.Button;
import android.widget.EditText;
import android.widget.ListAdapter;
import android.widget.ListView;
import android.widget.SimpleCursorAdapter;
import android.widget.Toast;

/* loaded from: classes.dex */
public class AccessControl3NotesActivity extends AppCompatActivity {
    /* JADX INFO: Access modifiers changed from: protected */
    @Override // android.support.v7.app.AppCompatActivity, android.support.v4.app.FragmentActivity, android.support.v4.app.BaseFragmentActivityD
    public void onCreate(Bundle savedInstanceState) {
        super.onCreate(savedInstanceState);
        setContentView(R.layout.activity_access_control3_notes);
    }

    public void accessNotes(View view) {
        EditText pinTxt = (EditText) findViewById(R.id.aci3notesPinText);
        Button abutton = (Button) findViewById(R.id.aci3naccessbutton);
        SharedPreferences spref = PreferenceManager.getDefaultSharedPreferences(this);
        String pin = spref.getString(getString(R.string.pkey), "");
        String userpin = pinTxt.getText().toString();
        if (userpin.equals(pin)) {
            ListView lview = (ListView) findViewById(R.id.aci3nlistView);
            Cursor cr = getContentResolver().query(NotesProvider.CONTENT_URI, new String[]{"_id", "title", "note"}, null, null, null);
            String[] columns = {"title", "note"};
            int[] fields = {R.id.title_entry, R.id.note_entry};
            SimpleCursorAdapter adapter = new SimpleCursorAdapter(this, R.layout.notes_entry, cr, columns, fields, 0);
            lview.setAdapter((ListAdapter) adapter);
            pinTxt.setVisibility(4);
            abutton.setVisibility(4);
            return;
        }
        Toast.makeText(this, "Please Enter a valid pin!", 0).show();
    }
}
```

Figure 6.81 AccessControl3NotesActivity source code.

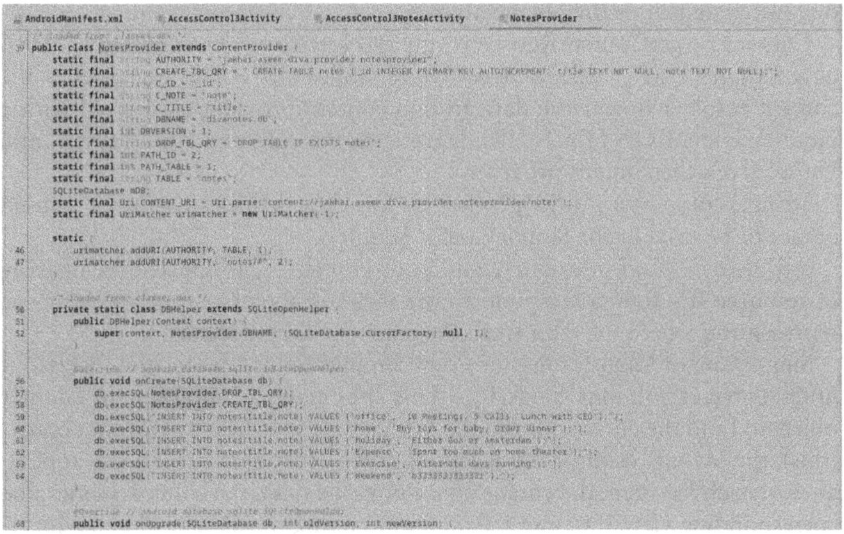

```
AndroidManifest.xml    AccessControl3Activity    AccessControl3NotesActivity    NotesProvider
/* loaded from: classes.dex */
public class NotesProvider extends ContentProvider {
    static final String AUTHORITY = "jakhar.aseem.diva.provider.notesprovider";
    static final String CREATE_TBL_QRY = " CREATE TABLE notes (_id INTEGER PRIMARY KEY AUTOINCREMENT, title TEXT NOT NULL, note TEXT NOT NULL);";
    static final String C_ID = "_id";
    static final String C_NOTE = "note";
    static final String C_TITLE = "title";
    static final String DBNAME = "divanotes.db";
    static final int DBVERSION = 1;
    static final String DROP_TBL_QRY = "DROP TABLE IF EXISTS notes";
    static final int PATH_ID = 2;
    static final int PATH_TABLE = 1;
    static final String TABLE = "notes";
    SQLiteDatabase mDB;
    static final Uri CONTENT_URI = Uri.parse("content://jakhar.aseem.diva.provider.notesprovider/notes");
    static final UriMatcher urimatcher = new UriMatcher(-1);

    static {
        urimatcher.addURI(AUTHORITY, TABLE, 1);
        urimatcher.addURI(AUTHORITY, "notes/#", 2);
    }

    /* loaded from: classes.dex */
    private static class DBHelper extends SQLiteOpenHelper {
        public DBHelper(Context context) {
            super(context, NotesProvider.DBNAME, (SQLiteDatabase.CursorFactory) null, 1);
        }

        @Override // android.database.sqlite.SQLiteOpenHelper
        public void onCreate(SQLiteDatabase db) {
            db.execSQL(NotesProvider.DROP_TBL_QRY);
            db.execSQL(NotesProvider.CREATE_TBL_QRY);
            db.execSQL("INSERT INTO notes(title,note) VALUES ('office', '10 Meetings. 5 Calls. Lunch with CEO');");
            db.execSQL("INSERT INTO notes(title,note) VALUES ('home', 'Buy toys for baby, Order dinner');");
            db.execSQL("INSERT INTO notes(title,note) VALUES ('holiday', 'Either Goa or Amsterdam');");
            db.execSQL("INSERT INTO notes(title,note) VALUES ('Expense', 'Spent too much on home theater');");
            db.execSQL("INSERT INTO notes(title,note) VALUES ('Exercise', 'Alternate days running');");
            db.execSQL("INSERT INTO notes(title,note) VALUES ('Weekend', 'b333333333333');");
        }

        @Override // android.database.sqlite.SQLiteOpenHelper
        public void onUpgrade(SQLiteDatabase db, int oldVersion, int newVersion) {
```

Figure 6.82 NotesProvider activity source code.

```
generic_x86:/ $ content query --uri content://jakhar.aseem.diva.provider.notesprovider/notes
Row: 0 _id=5, title=Exercise, note=Alternate days running
Row: 1 _id=4, title=Expense, note=Spent too much on home theater
Row: 2 _id=6, title=Weekend, note=b333333333333r
Row: 3 _id=3, title=holiday, note=Either Goa or Amsterdam
Row: 4 _id=2, title=home, note=Buy toys for baby, Order dinner
Row: 5 _id=1, title=office, note=10 Meetings. 5 Calls. Lunch with CEO
generic_x86:/ $ ▌
```

Figure 6.83 Accessing the notes by firing the content provider directly.

SharedPreferences spref = PreferenceManager.getDefaultSharedPreferences(this); – This line retrieves the default SharedPreferences instance using PreferenceManager in the current context (this).

String pin = spref.getString(getString(R.string.pkey), "“); – This line retrieves the PIN stored in the shared preferences using the key obtained from the string resource with the name pkey. If the key does not exist, an empty string is returned.

String userpin = pinTxt.getText().toString(); – This line retrieves the text entered in the pinTxt EditText view and converts it to a String by calling the toString method.

if (userpin.equals(pin)) {. . .} – This line checks if the userpin string is equal to the pin string retrieved from the shared preferences. If they are equal, the user has entered the correct PIN.

ListView lview = (ListView) findViewById(R.id.aci3nlistView); – This line retrieves the ListView view with the ID aci3nlistView from the current layout and assigns it to the lview variable.

Cursor cr = getContentResolver().query(NotesProvider.CONTENT_URI, new String[]{"_id", "title", "note"}, null, null, null); – This line queries the content resolver to retrieve data from a content provider associated with a NotesProvider.CONTENT_URI. It retrieves the columns "_id," "title,", and "note" from the content provider.

String[] columns = {"title", "note"}; – This line defines an array of column names to be used in the SimpleCursorAdapter.

int[] fields = {R.id.title_entry, R.id.note_entry}; – This line defines an array of resource IDs that correspond to the views in the notes_entry layout file, representing where the data from the cursor will be bound.

SimpleCursorAdapter adapter = new SimpleCursorAdapter(this, R.layout.notes_entry, cr, columns, fields, 0); – This line creates a SimpleCursorAdapter object to bind the cursor data to the ListView. It specifies the current context (this), the layout resource file R.layout.notes_entry that defines the appearance of each list item, the cursor containing the data, the column names, the corresponding view IDs, and a flag (0) specifying no additional behavior.

lview.setAdapter((ListAdapter) adapter); – This line sets the SimpleCursorAdapter as the adapter for the ListView, which will display the data retrieved from the cursor.

pinTxt.setVisibility(4); – This line sets the visibility of the pinTxt EditText view to View.INVISIBLE (4), effectively hiding it.

abutton.setVisibility(4); – This line sets the visibility of the abutton Button view to View.INVISIBLE (4), effectively hiding it.

Toast.makeText(this, "Please Enter a valid pin!", 0).show(); – This line displays a short-duration Toast message with the text "Please Enter a valid pin!" to indicate that the entered PIN is incorrect.

In summary, the AccessControl3NotesActivity class represents an activity in an Android application. The onCreate method sets the content view for the activity, and the accessNotes method handles the logic related to accessing notes based on a user-entered PIN. If the PIN is correct, it queries a content provider for notes data and displays it in a ListView. If the PIN is incorrect, it shows a Toast message informing the user to enter a valid PIN.

Now, let's check out the NotesProvider activity.

So, to explain the code to you the snippet includes a nested class named DBHelper, which extends SQLiteOpenHelper. This class is responsible for managing the creation and upgrade of an SQLite database used by a NotesProvider class. Here's a breakdown of the code:

public DBHelper(Context context) – this is the constructor of the DBHelper class. It takes a Context object as a parameter and calls the superclass constructor (SQLiteOpenHelper) with the database name obtained from NotesProvider.DBNAME and a null value for the CursorFactory. The last argument specifies the database version, which is set to 1.

public void onCreate(SQLiteDatabase db) – this method is called when the database needs to be created. It receives an instance of the SQLiteDatabase class, representing the database to be created. Within this method, several SQL statements are executed using the execSQL method on the db object.

The executed SQL (Structured Query Language) statements include dropping an existing table (NotesProvider.DROP_TBL_QRY), creating a new table (NotesProvider.CREATE_TBL_QRY), and inserting sample data into the table using several INSERT INTO statements.

public void onUpgrade(SQLiteDatabase db, int oldVersion, int newVersion) – this method is called when the database needs to be upgraded, typically when the database version changes. In this implementation, the method simply calls onCreate(db), which drops and recreates the entire database, effectively discarding any existing data.

In summary, the DBHelper class manages the creation and upgrade of the SQLite database used by the NotesProvider. The onCreate method sets up the initial database structure and inserts sample data, while the onUpgrade method handles the process of upgrading the database by recreating it from scratch.

So, we can comprehend that the private notes are being stored in an SQL database.

So, if you remember, in AndroidManifest.xml we saw that there was a content provider that was being implemented by the NotesProvider. So, if we go and check the NotesProvider, we see that there's a URI that points to that content provider.

So, by using content utility inside the Android shell, we can access the notes. Let's see.

Step #3: Go to the terminal and type: content query –uri content://jakhar.aseem.diva.pro vider.notesprovider/notes.

Since the notes were stored in an SQL database, the result came in a tabular form. Hence, we were able to successfully exploit the vulnerability.

There are two more tasks left, which I want you to attempt on your own. Consider it as your homework and try to exploit the vulnerabilities. Don't be afraid to view my video guide on my YouTube channel if you get stuck. This video link will aid you in your quest if you find yourself lost. Let's go on to the pentesting phase's most tedious but crucial phase, that is the reporting phase. This stage is what distinguishes a black hat hacker from a white hat hacker.

Chapter 7

Report writing and final words

7.1 REPORTING PHASE

In the entire process of locating and correcting vulnerabilities in a system, the penetration testing reporting phase is a crucial stage. The report's goal is to clearly, succinctly, and practically describe the results and suggestions of the penetration testing engagement. The following components must be included in a normal penetration testing report:

- **Executive Summary:** This part needs to include a high-level overview of the testing's goals, scope, and methodology in addition to a rundown of its most important conclusions and suggestions. It should be expressed in plain terms that nontechnical stakeholders may easily comprehend.
- **Testing Objectives and Scope:** A thorough explanation of the engagement's testing objectives and scope should be provided in this section. It should contain facts about the software versions, the systems and networks examined, and any other pertinent information.
- **Process:** A thorough explanation of the testing methodology, including the tools and methods utilized, should be included in this section. The timetable for the testing, deadlines, and any obstacles or limits that arose throughout the engagement should also be included.
- **Findings:** A thorough analysis of the vulnerabilities found during the testing process should be provided in this section. The degree of each vulnerability should be described, along with remediation or mitigation measures that might be implemented. Screenshots and any pertinent material should be included in the report to support the conclusions.
- **Recommendations:** This section needs to provide suggestions for enhancing the system's security. In addition to recommended practices for system and network security, this can also include suggestions for particular security measures that might be put into place.
- **Appendices:** This part should contain any further technical details, such as step-by-step instructions for recreating the vulnerabilities found, network schematics, and packet grabs. It should also contain

DOI: 10.1201/9781003450474-7

any unprocessed information gathered during the testing period, such as packet captures and log files.

The important conclusions and suggested actions should be summarized in this part, along with a request for the stakeholders to act on the conclusions and suggestions.

Stakeholders will have a clear knowledge of the vulnerabilities found and the measures that may be taken to mitigate them thanks to a well-written, thorough penetration testing report. It should be written in an easy-to-understand style that is clear and simple, utilizing technical jargon only when absolutely essential. It should also be presented in a way that is aesthetically appealing so that the main conclusions and suggestions may be communicated more effectively.

Along with the aforementioned components, it's crucial to make sure the report is provided on time and to follow up with the stakeholders to make sure the suggested corrective actions have been done. As new vulnerabilities and threats are found and as the system ages, the report should be frequently evaluated and updated.

In its entirety, the penetration testing reporting phase is an essential element in the overall security process and should be accorded the same amount of priority as the testing phase itself. The report should be thorough, actionable, and simple to comprehend. It should serve as the foundation for continuing security efforts to strengthen the system's security posture.

7.2 FINAL WORDS

All of these things seem fantastic, but consistency is what a beginner needs. Android pentesting itself is a complex field, and one needs to stay aware and updated about the newer updates and releases of Android security measures. As a beginner, it is suggested to get a good understanding of the basics of Android and slowly proceed toward advanced topics.

To learn Android pentesting, you can follow these steps:

- Learn the basics of Android OS and its architecture.
- Learn Java programming, as it is the primary programming language used for Android development.
- Learn the Android SDK and how to use it to create and debug Android apps.
- Learn about Android app components such as activities, services, broadcast receivers, and content providers.
- Learn about Android security mechanisms, such as permissions and encryption.
- Learn about mobile pentesting tools, such as Burp Suite, OWASP ZAP, and adb.

- Learn about mobile pentesting methodologies, such as black-box testing and white-box testing.
- Practice pentesting on real-world Android apps and look for vulnerabilities.
- Read articles, blog posts, and watch tutorials to stay updated with the latest Android security issues and patches.
- Join online communities, such as forums and social media groups, to connect with other security professionals and share knowledge.

It is important to note that Android pentesting is a complex and evolving field, and you will need to invest a significant amount of time and effort to become proficient in it.

This book aims to light up the interest of Android Pentesting among security enthusiasts. So, if you think you are lost, no worries. Below listed are the topics which you should be knowing in order to become a good Android pentester.

This book seeks to pique the attention of security enthusiasts in Android pentesting. So, if you believe you're lost, don't worry. The subjects that you need to be familiar with in order to become a proficient Android pentester are given below.

7.2.1 Basics of Android pentesting

- Understanding Android architecture and how it differs from the traditional desktop or server-based systems.
- Learning about Android security mechanisms, such as the Android security model, permissions, and the Android Sandbox.
- Familiarizing yourself with the Android SDK and tools, such as Android Studio, adb, and the Android emulator.
- Understanding the process of rooting and jailbreaking Android devices, and the security implications of doing so.
- Familiarizing yourself with mobile application testing methodologies, such as the OWASP Mobile Application Testing Guide (MATG).
- Understanding how to use mobile forensics tools to extract data from Android devices.

7.2.2 Advanced Topics of Android Pentesting

- Reverse engineering Android apps to understand their internal workings and identify vulnerabilities.
- Exploiting Android vulnerabilities such as those related to the Android OS, app components, and communication protocols.
- Analyzing and detecting malicious apps and malware.
- Understanding the security implications of popular third-party libraries and frameworks used in Android development.

- Understanding the security implications of newer features in Android, such as Android's security-enhanced SE for Android (SEAndroid), and the Android Application Sandbox.
- Performing advanced mobile forensics techniques to extract data from Android devices.
- Understanding how to test for vulnerabilities in Android-specific technologies such as NFC, Wi-Fi Direct, and Bluetooth.

These are the basic and advanced topics of Android pentesting, the resources, and tools for these topics are also available online which can help you to learn and practice the same.

As a pentester, you should be skilled not only in technical abilities but also in presentation and communication skills. In addition, as a pentester, you will spend the majority of your time reading documents rather than inputting commands in the black terminal. As a result, reading at least 10 pages every day is a sensible move. So, a few books (not sponsored) that I would recommend to help you build your reading habits and become a better Android pentester are as follows:

- *Mobile Pentesting for Beginners* **by Sabyasachi Paul** – This book provides a hands-on journey and guidance and gives a good overview about Android internal, security mechanisms, underlying architecture, and few vulnerabilities.
- *Android Hacker's Handbook* **by Joshua J. Drake et al.** – This book provides a comprehensive overview of Android security, including topics such as Android architecture, rooting, and malware analysis.
- *The Android Security Cookbook* **by Keith Makan** – This book provides practical guidance on how to secure Android apps and devices, as well as on how to test for vulnerabilities.
- *Android Security Internals: An In-Depth Guide to Android's Security Architecture* **by Nikolay Elenkov** – This book provides a detailed review at Android's security architecture, including the security mechanisms built into the operating system, and how they can be used to secure apps and devices.
- *OWASP Mobile Security Testing Guide (MSTG)* – The OWASP Mobile Security Testing Guide (MSTG) is a comprehensive guide to testing the security of mobile apps. It covers the entire testing process, including planning, reconnaissance, and testing for specific vulnerabilities.
- *SANS Institute's Mobile Device Security* **course** – SANS Institute's Mobile Device Security course provides a comprehensive introduction to mobile device security, including Android pentesting.
- *Practical Mobile Forensics* **by Heather Mahalik** – This book provides an in-depth look at mobile forensics, including the tools and techniques used to extract data from Android devices.

- *Android Pentest* **by Ajin Abraham** – This book is a comprehensive guide to Android application penetration testing. It covers the basics of Android security and goes on to cover the tools and techniques used in testing Android apps.

These resources will provide you with a good understanding of Android security and how to test vulnerabilities. I suggest you to start with the basics and then move to the advanced topics.

Anyway, we've reached the end of this book, but the journey of Android pentesting is far from over. I wish all the readers the best of luck on their future Android pentesting adventures. I hope you found this book useful and learned a lot of new things from it. If you have any questions or concerns, please contact me via social media. In the online community, I'm known as h0tPlug1n. So, that's all for now, and I'll see you all soon on another book. Adios!

Index

For Product Safety Concerns and Information please contact our EU representative GPSR@taylorandfrancis.com
Taylor & Francis Verlag GmbH, Kaufingerstraße 24, 80331 München, Germany

www.ingramcontent.com/pod-product-compliance
Lightning Source LLC
Chambersburg PA
CBHW070722220326
41598CB00024BA/3261

* 9 7 8 1 0 3 2 5 8 5 1 7 8 *